In Search of the Person

In Search of the Person

Philosophical Explorations in Cognitive Science

Michael A. Arbib

The University of Massachusetts Press

Amherst, 1985

Printed in the United States of America
LC 85-14152
ISBN 0-87023-499-4 cloth; 500-1 paper

Designed by Barbara Werden
Drawings by June Gaeke
Set In Linotron Trump Medieval at G&S Typesetters, Inc.
Printed and bound by Cushing-Malloy, Inc.

Library of Congress Cataloging-in-Publication Data

Arbib, Michael A.
In search of the person.

Bibliography: p.
Includes index.
1. Artificial intelligence. 2. Cognition.
3. Neuropsychology. I. Title.
Q335.A73 1985 006.3 85-14152
ISBN 0-87023-499-4 (alk. paper)
ISBN 0-87023-500-1 (pbk. : alk. paper)

This volume is dedicated to
Mary Hesse
in gratitude for
three years of lively debate
and to
Rolando Lara and Elena Sandoval
for all that was
and all that might have been

Preface

FOR THE LAST twenty-five years, I have been trying to understand both the workings of brains and new approaches to the design of intelligent machines. Such concerns often led me to ponder philosophical questions about the nature of mind and knowledge, but more narrowly scientific concerns left too little time to pursue them. However, when the University of Edinburgh invited me to deliver the Gifford Lectures I accepted with alacrity, not only because of the great honor involved, but also because I knew that the responsibility of preparing these lectures (I was given three years' formal notice) would force me to at last find time for the serious study of philosophy.

But this book is not the serious tome that such study led me to produce. That tome is *The Construction of Reality* (Cambridge University Press, 1986) which goes far beyond the scope of the Giffords. It was written with Mary Hesse, with whom I was invited to share the ten lectures in November 1983, and developed as we strove to understand our points of agreement and disagreement for Edinburgh. The present book grew from a series of five "Post Gifford" lectures given at the University of Massachusetts, under the auspices of the Institute for Advanced Study in the Humanities, designed to share some of the ideas from Edinburgh with friends, colleagues, and students. It is self-contained, and yet will, I hope, invite a number of readers to go on to read the more detailed and far-reaching account in *The Construction of Reality*.

The "Interview by Way of Introduction" opens the book by placing my philosophical explorations in the context of an ongoing concern with cognitive science in general and brain theory in particular.

Chapter 1 then addresses the question: To what extent can cognitive science (including artificial intelligence) give an adequate account of the person? I shall suggest that there are no limits to artificial intelligence as such, but still argue that there is much that is specifically human about our intelligence which depends both on our bodies and on our interactions within society. An intelligent machine must learn, and some element of logical inconsistency is inescapable in any system that makes intelligent decisions while interacting with a complex and changing world. This answers critics who see Gödel's Incompleteness Theorem as proving the impossibility of machine intelligence.

A central theme of this book is to build on Piaget's study of assimilation and accommodation in the child and my own work in brain theory and artificial intelligence to develop a *schema theory* which will link cognitive science to the study of persons in society. I seek to show how the hundreds of thousands of schemas in a single brain may cohere to constitute a single person, thus setting the stage for our discussion of freedom of the individual. Schemas provide the bridge between the unitary self and the billions-fold complexity of the activity of the neurons of the brain.

Chapter 2 outlines a schema-theoretic account of language as a socially constructed reality, while showing how that reality is embodied in the schemas of individual members of the language community, presenting a neo-Piagetian view of language acquisition. I explore both the commonality and individuality of the schemas held by different people. This account supports a view of language as metaphor, which I show to be necessary for philosophy of science as well as being central to the study of religious language and myth.

Freud provides a link from the brain-theoretic dimension of schema theory to what might seem beyond the reach of science, namely a concern with myth. Freud started as a neurologist, yet came to ascribe reality to the Oedipus myth as a causal factor in both ontogeny and phylogeny. In chapter 3, I offer a schema-theoretic account of consciousness and then compare this with Freud's theory to highlight the contributions of schema theory and also the extent to which it has not yet come to terms with the "darkness of the soul," as seen in those phenomena which led Freud to his theory of repression. I then explore Freud's views on religion both for his theory of projection—God as construct—and for his pessimistic view of civilization.

The fourth chapter introduces hermeneutics (the science of interpretation) to let us further explore continuities between the natural sciences and other symbolic systems of knowledge. After examining Marx's views on ideology, I present an interpretive theory of society, consistent with my view of language, in which social forces are rooted in the coherence of schemas held by individuals within a society, while social change is rooted in the discordance between individual schemas and social commonalities. Contrary to Habermas' notion of an ideal speech situation, I argue that any "acceptable" community will be characterized by pluralism rather than consensus.

In the final chapter, I turn to the question of human freedom, starting from the observation that cognitive science seeks mechanistic explanations for all human mental processes, including, in principle, the most complex and specifically "human"—intelligence, creativity, social communication, free will, and moral and religious belief. Does this destroy all that is valuable in the concept of the "human"? Recapitulating the debate between Mary Hesse and myself in our Gifford Lectures, I contrast the approach of the "voluntarist" and the "decisionist," respectively. The "voluntarist" argues that *if* we are to retain a common understanding of freedom as real choice, carrying responsibility and meriting praise or blame, *then* something beyond cognitive science, indeed beyond naturalism, is required. The "decisionist," on the other hand, espouses a reinterpretation of the concept of freedom as it has been understood in most religious traditions. On the basis of this view I present secular schemas which offer a reading of the human condition as lying wholly within the spatiotemporal realm, with no appeal to God or to a "voluntarist" free will. Such schemas would be in constant flux, yet maintain their ability to embody a multilevel reality encompassing persons and society as well as things.

No brief volume can fully treat all these topics, yet its very brevity may invite the reader to see these issues in relationships that have hitherto been obscured. This book should interest general readers who want to learn something of the ABCs of artificial intelligence, brain theory, and cognitive science, presented in a spirit of philosophical concern with the human condition, rather than in its hard-edged technological detail. But I also hope that it will be read by specialists in one or more of the ABC sciences who, like me, love their science and yet welcome the all-too-rare opportunity to ponder its broader implications, and the web of understanding of which it is a part.

MICHAEL A. ARBIB

Amherst, Massachusetts
August 1984

Acknowledgments

THERE ARE many acknowledgments to be made. Donald Michie and Barry Richards first supported the idea that I might give the Gifford Lectures; Jules Chametzky offered the support of the Institute for Advanced Study in the Humanities of the University of Massachusetts at Amherst both before and after my month in Edinburgh. My collaboration with Mary Hesse played a vital and stimulating role in the development of the ideas that pervade the present volume; and I should also pay tribute here to my late friend, Bernate Unger, who in reading early chapters of the enterprise, emphasized the lived experience of the person. My thanks to Darlene Freedman and Judy Rose for typing the manuscript. June Gaeke attended the "Post Gifford" lectures and this led to the drawings for this book.

In the summer of 1983, Rolando Lara—my closest colleague in brain theory and a very good friend—invited me to spend a week in Mexico to rehearse the Gifford Lectures, while staying at the Cuernavaca home of him and his beloved Elena Sandoval, a colleague and neurochemist at Universidad Nacional Autonoma de Mexico. During that week, Maria Pia Lara—Rolando's sister and a science journalist with *Naturaleza* magazine—conducted the interview which serves as introduction to the present volume. In one of those tragic accidents that reminds one of how fragile our hopes and dreams can be, Rolando, Elena, and Willi Borchers (a friend and colleague visiting UNAM at the time) were killed on 19 January 1985 when a truck ran into their car on a road near Cuernavaca. How I wish that we could have once again debated the ideas in this book.

Contents

In Search of the Person

An Interview by Way of Introduction

Lara: About your work, a little introduction.*

Arbib: I started life as a mathematician, but as an undergraduate I was introduced to Norbert Wiener's book on cybernetics and in my summers I learned to program computers. When I finished my undergraduate work the place to go was MIT because that was the center of cybernetics. At MIT I learned about artificial intelligence from Marvin Minsky, about cybernetics from Norbert Wiener and Warren McCulloch, and about information theory from Claude Shannon. Halfway through my Ph.D. I went back to Australia, and to earn money for the term I was there I gave a course in what I had been learning about—modeling brains and making mathematical theories of machines. The course was called "Brains, Machines, and Mathematics,"

* An edited transcript of an interview of the author by Maria Pia Lara, Mexico, August 1983. The interview was commissioned by, and is used here with the permission of, the Centro Universitario de Communicación de la Ciencia of the Universidad Nacional Autonoma de Mexico. A shorter version, in Spanish translation, was published in their magazine *Naturaleza*.

and the notes became a book. The rest of my career for the last twenty years has been working out that theme of brains, machines, and mathematics.

I've done some work in purely mathematical systems, then later mathematical theory of computers and programming. I've also worked in artificial intelligence—how to make computers understand certain aspects of language, or how to give them a little bit of visual ability, and how to make computers control the behavior of a robot. Perhaps my most important interest, though, has been trying to use this mathematical knowledge, this knowledge of computers, and these ideas from artificial intelligence, to understand the brain. For many years now I've been trying to introduce a theoretical approach to the brain. It's still the case at the moment that most people working on the brain are experimentalists who think that the only good research is to be conducted in the laboratory. But there are quite a few who realize that to understand the complexity of behavior requires not only a special type of experiment on how a human or an animal behaves but also the use of the computer in two different ways: as a tool—so that we can explain a complex problem, write a computer program, and use that program to understand what these ideas really mean—and as a metaphor.

I wrote a book called *The Metaphorical Brain*. The brain is a bit like a computer, and a lot of the research has been to understand how to learn from thinking about computers without reducing the brain to nothing more than a computer. A brain is attached to a body; that's different from a computer. A brain has billions of neurons, all active at the same time, thousands of regions. So instead of thinking of the normal computer—which just carries out one instruction at a time, doing a problem somebody else has told it to solve—the problem for brain theory is to understand how a brain in a body with many, many things happening at the same time can see, can act, can learn.

The Nature of Brain Theory

Lara: I want you to explain the concept of artificial intelligence, then brain theory.

Arbib: We can look at a person behaving, we can look at the way they look at something and recognize what it is, the way they hear a question and answer it correctly, and we can ask how they do it. One way of answering this would be a psychological explanation which would say something about certain memories and being able to retrieve certain things from memory. An artificial intelligence scholar might say, "That's all very well, but you haven't said how you would store information in memory, how the pieces of information would be connected to each other, what process determines what you want to retrieve from memory." He would say, "I don't care very much about what happens inside the human, I want to know how to make a machine that can answer questions for me, or can do some pattern recognition for me. So I'm going to take your psychological ideas as a starting point, but my real aim is to write a useful program. I may store information in the memory of my computer, but that doesn't mean that I think that the wiring of the brain looks like the wiring of the computer. It's just that I have a convenient way of storing it. I might use a very quick search with the computer to find something, but I think probably the mind uses something different." That's the first level, that's *artificial intelligence*.

The next level is *cognitive psychology*. Here people say, "I think you're right that we need to use this precise language of what ways information is stored, what are the precise procedures to get information from memory, and so on, if we're really to have a good model. But I want to explain how humans do it; so a cognitive model must be successful in the way a human is. It's got to be slower at things humans are slow at; it's got to make the sort of errors a human makes. So I'm no longer trying to make the program that's most efficient to get a machine to do a job for me; I'm how trying to get the program that makes the machine behave in a way that helps me understand how the human does it."

But *brain theory* imposes even further constraints. The brain theorist says, "That's all very well; you've given me an idea of how to write a program that behaves from the outside in terms of what is difficult or what the errors are or what the speed is. The program may predict the results of experiments on human behavior, but it's not a brain model." The brain theorist says, "If I'm going to make a model of how a human behaves or how an animal behaves, I want it not simply to yield the same behavior from the outside, of how stimulus yields response, or how an animal explores a new situation. I want to have the model help me understand information about how brain damage changes behavior. A neurophysiologist can put an electrode in the brain and check what a few cells are doing in different behaviors. I want my model to show how different cells have different roles in carrying out that behavior so that I can begin to predict what new cells will do when the animal does something different."

One can be a brain scientist, a neuroscientist, either as an experimentalist going in to do more observations of how different parts of the brain work in different situations, or as a theorist who says, "Now there are many experiments. Can we begin to use mathematics to unify these different experiments to see how one set of ideas about certain circuits in the brain can explain many different experimental results, or perhaps lead us to new hypotheses which can lead to new experiments?" Such a person would use computer simulation and mathematical analysis, but the important point is that his model is not just a program that does a job; his model is a program or a mathematical model that represents how different pieces of the brain interact with each other to explain the overall behavior of the animal or the human.

Lara: Like a theory of systems?

Arbib: It's a theory of systems in which the subsystems have to be mapped to different areas of the brain. It's not enough to find systems with the same input/output behavior.

Lara: Do you think that using cybernetics and a mathematical approach could really be the way to learn what the mind can do; inferring the mind's properties by purely rational analysis, as scientists in the seventeenth century tried to do?

Arbib: Noam Chomsky thinks we can understand the mind by rational analysis, but I think what brain theory is saying is that while we can learn from Chomsky, while we can learn from artificial intelligence, to understand the human mind properly we also have to do a great deal of experimental work. Part of it is just looking at how people do tasks: What mistakes do they make? What do they do quickly? What gives them trouble? What can't they do at all? Looking at brain damage, what happens to a person who has something wrong with one part of his brain or another part? It also involves looking at animals to find human analogies. We have two advantages over the pure rationalists: One is that we have a whole body of experimental techniques. We can now examine single cells in the brain. We begin to be able to follow patches of activity in the brain of an awake human. But also with the computer we have the ability to make theories that are not just what we can write in a few pages of print, but ones that we can express dynamically—a computer can say what happens in this circumstance, what happens in that circumstance—and try out many alternatives. For example, today if we make a model of the brain we don't have to just make one model and make an argument with a little hand waving as to why it seems plausible; we can make that model in such a way that we can put it in the computer and try out different hypotheses within that same program. So we can test: Is this part of the brain connected to that part of the brain? Does it seem that this is represented here or there?

Lara: I also think that mathematics has again become a vital element for rational explanation of the mind and brain now that theory has gained a cybernetic theory of systems. A lot of fields do need mathematics.

Arbib: The important point to bear in mind, though, is that mathematics is a very varied topic. If you go to university, you may learn some basic calculus, geometry, algebra, and you get the impression that there's a nice set of topics that constitutes mathematics. In fact there are many new and different areas. When we come into a new area like brain theory, we take some ideas from the study of chemical systems, some from the theory of games, but as we get further into the problems of the brain we have to start making new mathematics, finding new techniques to help us answer new questions. Mathematics is not a complete body of knowledge; rather it's a way of using symbols as a shorthand that can be manipulated to try out different alternatives. The big discovery, I think, is that for very complex systems sometimes we can use mathematics to formulate the problem but cannot use it to solve the problem. We can use it essentially to say, "Here are a number of alternatives," but whereas in geometry we might say, "Here is the proof that this is the right alternative," often in modern mathematics, having found those alternatives we have to design computer programs that can search these many, many alternatives.

Lara: Brain research has been mainly experimental. Do you feel that with the use of mathematics it could go further?

Arbib: Yes. The problem is that the brain is so complicated, has so many different aspects, that it's not going to yield in the same way as astronomy does. With astronomy a Newton could really take the data on how the planets move and Galileo's experiments on a ball rolling and so on to derive a set of first principles for the subject. But different animals have very different brains; brains have millions and even billions of cells in them; there are many interesting details below the cellular level, at the cellular level, and concerning how a particular region works. As a result there's going to be an immense diversity of facts to be discovered by pure experiment. This means that many experimentalists will in fact make great contributions without explicit theory. Some such researchers then say, "Well, I don't need theory so why should we need it at all in neuroscience." By contrast astronomy was always a calculational science, so there is a respect for theory there.

But there begin to be a number of successes in brain theory. At the level of how one neuron sends a message to another neuron, there's the work of Hodgkin and Huxley. They won a Nobel prize for coming up with a differential equation description of impulse propagation. Basic things like breathing or walking involve rhythms, and there begin to be good mathematical descriptions of rhythms controlled by neural networks. There are a number of us, including Rolando Lara of the Autonomous University of Mexico, who are working at a higher level where, instead of worrying about the properties of the single cell or a few cells, we try to make the connection with behavior. "If we have a behaving animal, what is the basic program (in artificial intelligence terms) that would make it work?" Because the program is in a brain and not a computer we think about it as being a parallel program with many things happening at the same time, not a serial program doing one thing at a time. Because we are worried about brains, we don't simply say, "Here is a program which can respond to a visual stimulus as a frog would," for example. We say, "How can we correlate this with beautiful experiments about activity in one part of the brain, or the effects of removing another part of the brain?" So we begin to have a way of using mathematics and the computer to think not only about how one neuron at a time responds when being measured by the experimentalist, or the anatomy which is static, but tells us about many neurons. But to begin to put this together, to give us ways of thinking about what happens when many, many neurons are all active at the same time sending messages back to each other, making complicated patterns in the brain. I don't believe that this can be done by experiment alone; it's going to need the computer to help the experimentalist keep track. Experimentalists have already accepted the computer as an invaluable tool for collecting data. The next step is to convince them that it's also an invaluable tool for using mathematics to understand how behavior really is mediated by the brain.

Lara: Do you think the analogy with the computer will become the central guide for understanding the brain?

Arbib: I think I'll have to answer that in two parts. The first point is that the computer is really many machines. The computer by itself does nothing; it has to be given a program. You can give a computer one program, and it's a machine for retrieving information; another program and it's a machine for doing the payroll of a company; give it yet another program and it's a machine for helping a student learn arithmetic. The analogy was not with the computer itself, but with the computer-with-a-program. The analogies are not going to be with just any programs, but with the programs written in artificial intelligence—those programs in which people are trying to program a computer to do the sort of things that we think it requires some intelligence to do. In some cases I think the parallels will be very close. For example, in the work in computer vision, in seeing how to take a picture and analyze it to see what objects are represented in the image, there is already very close interaction between brain research and computer research. For example, people in machine vision now accept ideas about parallelism that computer people used to reject and that only the brain theory people were using. Now they see that processing the millions of bits of information corresponding to the light falling on the retina requires great parallelism. The machine vision person must turn the idea of parallelism into an efficient program. By understanding that efficiency we can perhaps get better ideas about different strategies a brain might use and make new hypotheses about the brain.

If we take another area like language: It's very hard to think about language without computer models which use a lot of information about words and their meanings; information about context; information about grammar; and procedures to bring all that information together to allow us to understand each new sentence. The understanding we get from computational linguistics is more in helping us think precisely about cognitive processes than in giving us a picture of how different parts of the brain interact. It requires an extra step by somebody who understands both the brain and artificial intelligence to go back and forth between them, a sort of thesis-antithesis to make a new synthesis which suggests a parallel model of language function.

Beyond Chomsky and Piaget

Lara: You once conducted a seminar with the provocative title "Beyond Chomsky and Piaget." What do you have in common with Piaget and what are your differences? Do you have points in common with Chomsky, and what have you done in the study of language?

Arbib: Piaget offers a constructivist view of the mental development of the child which starts with sensorimotor skills. He tries to explain the basic abilities of the child starting from a few schemas, a few basic patterns of action, for grasping or suckling, and various other things. Then he has essentially a descriptive theory of the types of skills the child develops as he (or she) grows from infant to adult—going from very simple motor skills to more abstract skills in which the child begins to have a notion of an object in general rather than just interacting with particular things; to a situation in which he begins to use formal arguments; and finally to the stage where, when appropriate, he can use logical arguments to solve a problem by using abstract knowledge. At each of these stages, Piaget talks about the person as having certain schemas, and he talks about the way in which these schemas change in terms of assimilation and accommodation. The idea is that if you are in a new situation, you try to make sense of it in terms of the schemas you have; that's *assimilation*. You try to assimilate the situation to what you know. But from time to time you will find yourself in a situation where these schemas are inadequate; what you know will not suffice. In certain cases, you change the schemas, and Piaget calls that *accommodation*; the schemas accommodate to the data they cannot digest, cannot assimilate.

With all this I'm very sympathetic. Where we differ is that Piaget seems to talk as if the stages are predetermined steps in maturation; whereas I want to see the interaction of the schemas as being the determination, so that eventually the schemas will develop a "common style" and we can talk of a new stage. The schemas will change again through accommodation and eventually a new style will emerge, and this will be another stage. I like Piaget in that he describes phenomena at the level of both stages and schemas that I want to understand as part of my theory of the mind. Where I part from him is that I think he is more descriptive than explanatory; he does not give a system theory or a computational model of how schemas change.

Lara: Is it possible for you to give us a sketch of this theory of schemas? Does it have some postulates? Does it have some rules? Does it have a vocabulary? Are there some principles?

Arbib: Schema theory is not as well defined as the questions suggest. The concept of schema goes back in neurology, for example, to Henry Head who talked about the body schema. He was trying to make sense of those people who had parietal lesions and "lost" half their body, so that in getting dressed they would put the shirt on one side of the body and leave it off the other. They didn't think it was part of their body because it wasn't part of their body schema. Common sense would say that nobody can ignore half of his body. But, in fact, with a parietal lobe lesion on one side of the brain, the representation of the other side of the brain may be lost. The "opposite" phenomenon discussed in neurology is the "phantom limb" where the limb is gone and yet the patients can feel where it is and can feel the pain in it.

That was the first round of schemas within the modern tradition of brain theory and cognitive science and artificial intelligence. Then Frederic Bartlett, who was a student of Head's, published his book *Remembering* in 1932. He observed that people don't seem to remember in a photographic way. If you tell somebody something and ask him to repeat it he doesn't repeat the exact words, but offers a reconstruction, recalling certain aspects and then providing words to re-express them. There is a party game where you sit around a table and one person writes down a little story and whispers it in the ear of his neighbor who whispers it to the next person, and so on, until the last person writes it down as he hears it. Then you read the two and everybody bursts into laughter because the stories are so different. Here, then, is the notion of remembering not as a passive process but as a constructive process in terms of the schemas that each person brings to bear to that construction. That I think would be the next chapter in schema theory.

Then we go through a sequence of papers within cybernetics, starting perhaps with Kenneth Craik's 1943 paper on "The Nature of Explanation," which says that the job of the brain is to model the world, so that when you act it is because you have been able to simulate the effects of your action before you do it. From then on there is a whole series of papers by people like

Donald MacKay, Richard Gregory, Marvin Minsky and so on who published papers about the model of the world in the mind or in the brain. This idea that the brain models the world continues to be basic. Then we start artificial intelligence and some of these ideas become expressed in computer programs, and today there are various mutations of this schema concept within artificial intelligence: Minsky uses the word 'frames'; Roger Schank uses the word 'scripts'; there are 'semantic nets'; and so forth.

Piaget belongs to another stream in schema theory that perhaps belongs with Kant rather than to this Head-Bartlett-cybernetics/Craik line. One of his books is called *The Construction of Reality in the Child*, and he says that the child at any time is building up a set of reproducible skills, or schemas, and that these make up what he can do. When I talk about schema theory, I'm talking about my own brand, but building on all these contributions.

Lara: Is there any difference between schema theory and work in artificial intelligence?

Arbib: In artificial intelligence, there are various attempts to talk about how a computer program, an artificial intelligence, or a robot, would model its world. They draw on the same roots in the earlier work of cybernetics that I do in developing my schema theory. I think the difference is twofold: because I have been thinking about animals (this is where many of the good experiments on the brain are being conducted), I've probably tended to think much more about how schemas relate to the basic sensorimotor skills of perceiving a situation and deciding how to act. People in artificial intelligence, however, starting with the computer as a symbol manipulator, have tended to emphasize such symbol-manipulation problems as language processing, playing a game of chess, or proving theorems in logic. But I would suspect that in the long run we can see these as two partial views: one emphasizing what is easy to program on the computer for symbol manipulation, the other emphasizing what can be related to studies of the brain of an animal that solves problems of a much more living-in-the-world kind rather than an abstract symbol-manipulating kind. At the moment, the two are very different, but I think probably in the long run the full development of schema theory will look back to each of these efforts as part of the history of the unified theory that emerges.

Lara: From this point of view, then, we have to think that Piaget provides a useful approach to schemas and to integrate this approach into an account of brain mechanisms for ongoing behavior.

Arbib: In fact, once we start worrying about the details of behavior, the concept of schema has to change. For Piaget the schema tends to be the overall skill shown in the current situation, whereas my theory requires a whole variety of schemas even in a single situation. When you behave in a particular situation, you have to recognize many things—the people around the table, the table itself, where the margarita is—and this means that one has different schemas for recognizing the drink, the table, and the people. One also has ways for combining those schemas so that one can represent a totally novel situation and yet call upon the knowledge that one has to make sense of that situation. One of the most important concepts that I would claim as mine rather than Piaget's is that of the schema assemblage; that it is at any time a network of interacting schemas pulled together that represents the situation and gives you the knowledge to handle this new situation. For Piaget, the schema tends to be a much bigger general skill. I don't deny the importance of such things, but I have, as it were, a microanalysis of how the ability to recognize many different objects, the ability to carry out many different actions, must be integrated to give you a wide ability to handle novel situations in part, at least, of their complexity. Thus, there have to be mechanisms whereby schemas can be activated, pattern recognition routines. There have to be means whereby schemas can compete and cooperate with each other. And there have to be output routines, motor routines. Moreover, in most of my work I have the extra dimension of concern for brain mechanisms so that I don't stop at saying, "Here's a good description of a schema"; I want to say, "How could it occur in the brain?" Piaget is biological, but his biology is not that of brain. He makes analogies between embryology and mental development, and that's a different game.

Lara: Now about Chomsky.

Arbib: I said that Piaget gave a view of how a child's first experience develops more and more complex schemas so that, in particular, it eventually acquires the concept of objects and acquires the ability to name them and then acquires the ability to speak in bigger and bigger sentences, etc. For Piaget and for myself, one has the notion that it must require something rather complex in the brain to support these processes whereby simple schemas change into more complex schemas. But still it's a constructive process: through experience these things change. When Chomsky looks at language acquisition, it is very different. Let me parody Chomsky. When I read his book called *Reflections on Language* I was tempted to write a review of it called "The Three Faces of Chomsky." (You may remember *The Three Faces of Eve* about a woman with three different personalities.) There were three Chomskys in this book. First is the Chomsky who has given us a theory of grammar, and I think there is no doubt that he is one of the greatest contributors to our understanding of grammar.

The second Chomsky talks about the acquisition of language, making what I think are dogmatic statements that have no necessary relation to his theory of syntax. He says, "I have a theory of adult grammar, and it's very complicated. It seems to me amazing that in seven years a child could master all this, so I postulate that many of the constraints on grammatical structures are part of the genetic programming, the innate competence, of the child." For Chomsky language acquisition is not the acquisition of concepts and the ability to denote them with words and then to string these words together to represent more complex situations with, in a sense, the ability to live in the world driving the acquisition of language. For Chomsky it is a much more abstract process, almost purely syntactic, in which, with a very rich grammatical structure already in place, to learn Spanish or English requires a few triggering experiences rather than a lengthy period of constructing the language on the basis of experience. My own feeling is that Chomsky has not yet offered a convincing argument for this view of acquisition, and if we look at the data on how children acquire language, there are so many rich phenomena about how the child changes her language from week to week. In the nine weeks that Jane Hill spent studying a two-year-old child, this child went from usually speaking two words at a time to speaking three or four words at a time. This doesn't make any sense in terms of adult characteristics, but in terms of a computational neo-Piagetian view, Hill was able to make excellent sense of this week-by-week process of change. So while I respect Chomsky immensely for his theory of adult grammar, I believe he is wrong about acquisition. And I think it is unfortunate that the excellence and importance of his contribution to syntax has made many linguists respect his theory of acquisition as well. Chomsky's innateness theory is rationalist as distinct from empiricist, whereas Piaget's theory is a mixture—rationalist in the sense that certain mechanisms for schema change have to be built in, but empiricist in the sense that the way the schemas develop depends crucially on the experience of the child.

Chomsky has written many essays on political issues, and this is his "third face." To my mind, during the Vietnam War he was one of the saner voices. Currently in his discussion of issues, I think he has become too rigidly against the establishment and does not have the flexibility of discourse. The strange thing about this is that he claims his politics is strongly dependent on his other views. He feels that his theory of grammar forces his theory of innateness, and that this rationalist theory is necessary for his political theory because he says that if you accept an empiricist theory then humans could be brainwashed, trained to believe everything. So because he wants there to be basic human dignity, he rejects such plasticity, and he argues that this fits in with his rationalist view. He rejects the views on conditioning of people like B. F. Skinner. There are two responses to this: one is the flippant response that if we must have the genes that gave us Hitler and Stalin then it's hard to see that this innatist view is encouraging for human dignity. The other thing is that his arguments are, I think, just the arguments of a person responding to the situations in his society in a way that, despite his claims, does not depend upon his basic view about knowledge. This independence is very clearly demonstrated by the controversy over sociobiology.

Society and Religion

Lara: What is the relation between brain theory and sociobiology?

Arbib: E. O. Wilson in his book on sociobiology, and later books like *Promethean Fire* with Lumsden, takes a genetic view of culture. Let me contrast this with my own, rather preliminary, analysis of culture. I might try to give an analysis of culture in terms of schemas, but I would see the schemas that constitute culture as being the schemas acquired by an individual within a social milieu. Because there is interaction between persons within this community, they have many schemas in common and thus we can talk about a culture. But I would relate the concept of culture to a theory of how individuals acquire schemas. There are already many people who believe certain things; a child grows up in that community; he learns the language of that community; he learns the beliefs of that community; but his individual experience may lead him either to accept those beliefs or question them. My theory of the brain, my explanations of Piaget, would suggest that the constraints of the genetic structure of the brain are relatively mild, yet our brains are much more complex than are those of frog, say, so we have language, we have consciousness, we have the ability to criticize. But when it comes to the difference between Mexican culture and Spanish culture, between North America and Mexico, or between a culture that believes in infanticide and one which finds such a thing abhorrent, I see these as differences that arise in the history of a community of interacting people. I would not see any genetic difference, whereas the difference between a human and a chimpanzee in the ability to use language does go back to a genetic account.

Now the sociobiologists want to say that many of our cultural "units" are not schemas that the individual acquires as he grows up in a particular setting, but have themselves been selected for, and are genetically wired in, the brain. The sociobiologist wants to make far more of the schemas prewired than I would. However, sociobiologists have come under tremendous criticism from the left who find this whole notion of genetically prewired, determined culture as abhorrent. They want, after all, to be able to offer radical critiques of capitalism, and they don't want to accept the position that "it doesn't do you any good to criticize society because we are wired up to be capitalists, and nothing you can do will change it." Now we come to the amusing point: the social analysis of the critics of sociobiology is identical with the social analysis by Chomsky when it comes to the views of the day—a somewhat leftish antiestablishment view. But their views of the underlying nature of the mind is totally opposite because Chomsky wants to see so much wired into the brain to undergird language. In terms of epistemology, a Piagetian would be much closer to a critic of sociobiology, and Chomsky would be much closer to a supporter of sociobiology.

I think this is ironical as a comment on the intellectual life of North America, but also depressing in the sense that it suggests how far political views can be independent of epistemology. Well, that's not depressing; it just suggests that one of the beauties of the human mind is its very flexibility! It can support so many different languages and so many different cultural systems that the challenge is to understand the human brain not as something that will develop in a predestined way to perform specific actions, but rather as something that will, by interacting in a complex society, both assimilate something of that society and yet provide its own critique of that society, thus contributing to the continuing dynamic of culture and society.

Lara: Would you think that an approach from brain theory to social science would be Piagetian? Or could there be any other approach to social science?

Arbib: My own view is that science constitutes a network, that there are so many methods of scientific investigation, so many important questions, that there is no such single thing as science. Rather, there are many, many sciences. Brain theory is one of them. Psychology is another. Sociology is another. If you have an Anglo-Saxon definition of science, you restrict it to the "hard" sciences, the ones in which you can measure and find numbers and make careful numerical analysis. If you accept the continental notion, you have German *Wissenschaften* which include both the *Naturwissenschaften* of the physical sciences and the *Geisteswissenschaften* of the human sciences. This gives a more unified notion of systems of knowledge which ranges from physics to history and sociology.

It seems to me that you can't expect any one of these, at least in the foreseeable future, to be the anchor point for any other. We can't reduce everything to physics. We can't even reduce everything to brain theory! For a historian to say that an analysis of World War II must rest on a postmortem on Hitler's brain would be absurd. Yet I think that attempts to look at history in purely social terms, to see a purely historical determinism, also fail. To understand the history of the 30s and 40s, you have to understand Hitler as an individual, even though what Hitler did could not have happened without the Treaty of Versailles and its effect on the German psyche, or hyperinflation—topics which call for social and economic analysis. Social analysis might tell us that it was inevitable from 1923 on that there would be a world war initiated by Germany, but the particular individual pathology of Hitler surely is responsible for the form of the Nazi genocide.

Lara: Does that mean that brain theory involves only individual schemas? There is no such thing as social schemas?

Arbib: No. I would say that if you were going to look at history you might find that a very useful analysis would combine analysis of social forces with a psychological analysis of the individual psyches of some of the key actors. Neither pure social nor historical determinism nor pure psychology are enough. We have to go beyond either science, psychology or sociology and put them together. On the other hand, if I'm looking at a single person I might use the psychological analysis of Piaget or of Freud and learn much from these men, yet still find important things they don't understand. If we take brain theory and connect that, we can begin to make a better psychology. Alternatively, because I have the questions of psychology, I can make a better brain theory. My feeling is that psychology will change in interaction with schema theory and brain theory, that one's understanding of persons will grow by looking at them in dramatically stressful historical situations, and that as we understand how people behave in certain situations it will change our view of the social sciences. But at the moment I can't see a direct link. Brain theory and schema theory interact below the level of the person to help us understand the person, but our everyday experience as people helps our theory of the person, providing data that go back to refine the schema theory and the brain theory.

On the other hand, we have to accept the fact that social forces aggregate what people do, so that in some sense one has to look at social theory in terms of a collective theory of individuals. If you have 1,000 people, you can't give an exhaustive account of each one of them; so you give an overall historical or sociological or economic statement about them; yet to some extent that statement is not going to make sense unless you go back and learn a little bit more about how they interact as individuals. Social theory must also take into account that occasionally there are personalities in positions of power whose individual psyches become historically determinative, as in the case of Hitler. Study of the person can interact with a social analysis; but study of the person is enriched by the schema or brain level of analysis. And now, because our view of the person begins to be in part expressed in schema language, we can talk about social schemas in terms of how it is that the schemas in the heads of individuals may have enough in common that we can talk about these individuals as constituting a community, which reflects social forces rather than individual interactions.

Lara: What is the relation between ideology and schema?

Arbib: I would say that an ideology is expressed in the head of the individual as a set of high-level schemas which he has assimilated through being in a community where many people have similar schemas.

Lara: The relation between brain theory and religion?

Arbib: Historically, a major question for people who would ask about religion is whether there is a separate soul. You might ask, "Is there a God-reality separate from physical reality?" Then you might ask, "Is there a soul separate from the body? Is there a mind separate from the body?" These are intermediate levels because there are some people who will say there is just a physical reality and that the mind is just a property of our complex brain and bodies as we interact with other people. Other people say, "No, there is a mind separate from the brain, and understanding the brain will only explain a certain amount of what people do. Beyond that there is this separate mind." And such people would say that we have a free will of a kind that requires that there be something other than the body and brain and external interactions determining what we do. The more religious might then say, "No, it's not just a mind that exists while our bodies exist and that disappears at death. There is a soul, an eternal soul, and this soul has meaning in terms of a relation to God." At the moment brain research has almost nothing to say about these issues. We can explain certain properties of vision, of memory, of how we control movement, aspects of motivation. We certainly can't explain in terms of brain mechanisms how it is that a person can decide that Shakespeare is a better writer than Mickey Spillane. So we're certainly not in a position yet to say that the nature of the brain is such that it explains everything about the mind.

There is a famous series of lectures given in Edinburgh called the Gifford Lectures on Natural Theology: what can the study of nature tell us about religion? These were recently given by Sir John Eccles, who is one of the greatest contributors to neurophysiology, and he came out, rather dogmatically I think, in saying that the mind is separate from the brain. He believes his personality requires essentially that there be this Him in this Mind in a separate reality from spatiotemporal reality. I'll be giving the Gifford Lectures this year [1983], and I will be arguing with just as little evidence that the mind can be explained in terms of schemas interacting and that the interaction of schemas is embodied in the brain. Where Eccles appears to be dogmatic in his view that the mind must be beyond the brain, my position is somewhat more agnostic. I would say rather that at any time as a scientist I have certain problems that are already solved, certain problems that I've set myself to solve in the next few years, and then a general feeling about the course of the decades or centuries ahead, opinions which will probably be drastically different five years from now on the basis of what I do during that time. But, at least in terms of my current appreciation, I can see many problems about the mind that *in outline* I can understand how schema theory could solve. At the level of a five-year program, I say with enough confidence to stake my own work on it, that a number of the properties of the mind will be shown to be explicable in terms of schemas and brain. So, extrapolating, I say "I don't see any obstacle to any property of the human mind being defined in this way." This is my working hypothesis.

What is interesting about my Gifford Lectures is that usually one person gives them, but in this case they've invited two people to give them together. I'm giving them with Mary Hesse, a philosopher of science at Cambridge. She has looked at the way in which theories of physics, scientific theories change over time. This has led her to a view of epistemology very similar to my schema view, i.e., that you don't have a fixed knowledge of reality but rather a partial knowledge of reality, and that this can change with experience and sometimes will even change dramatically. In a person's life there may be a conversion or falling in love, in the life of science it might be something like the transition from Newtonian mechanics to relativity. We have been able to construct an epistemology, a theory of how people know their world, based on schema theory that we agree about. However, having done that, we disagree about the central issue of theology. Where I see the world in secular terms, Hesse shows that our epistemology is open to a coherent symbol system which embraces both modern science and belief in a nonspatiotemporal reality.

Lara: Art is also a way of knowing, just like religion. What can you tell us about it?

Arbib: It's certainly an important part of my life to enjoy good music—actually bad music too—or good art and beautiful landscapes. A society without art would be impoverished. I think the attempts of the Soviet Union to harness art to social realism were disastrous and diminished what it was to be a person. If you then ask, "Is there some platonic ideal that art expresses, is there some supreme art?" my answer, consistent with the others, would be, "No. What art is may vary dramatically from society to society, and consequently the notion of beauty is going to be different from society to society. I think there will always be a notion of beauty, but although many people may think of it as an absolute, I don't think it will be." But now we're in an area where not only does my brain theory tell me nothing, but to which I have not really given any serious analysis.

Lara: One day I will invite you to give some lectures about art so you will come and do that serious analysis. But let's return to your Gifford Lectures. Freud has suggested that religion is an illusion, but that it helps many people face the problems of life. Do you think this is enough of a reason to advocate religion?

Arbib: Following the rituals, living according to the provisions of religion, can indeed help us mute the discontents of civilization. But to believe on this account is a position which Freud would reject. Also I think it's a provision that Mary Hesse would reject because she would see it as smacking too much of an elite manipulating the masses in a cynical way. Marx talked about the opiate of the masses, so I think there is a denial of that particular course both by people within the religious tradition and by people outside the religious tradition.

The first question we must ask if we are in some sense to

accept Freud's view that religion is an illusion, is, what is this secular schema that people are to live by if they reject the "great schema" of the Bible? It's probably a somewhat stoic schema. It's not one that promises utopia; it's not one that says human lives will be free of pain; it's not one that says everyone will be able to go to their deaths happy in the knowledge of salvation, but at least it will be one that will try to see beyond the illusions. At the same time we must ask, with Mary Hesse, what are the claims of the Bible to forestall such rejection? How do people come to rejct Freud's claim that religion is an illusion, but instead see religion as defining human reality?

As background, we must look at the questions of hermeneutics, of interpretation and reality, of what it means for there to be a social schema. What are the standards whereby a social schema is tested, and in particular how does language enter into it? How does language help convey a social schema and contribute to its interpretation? My position is that the mediating reality in all our discussion here is person reality and that brain reality or physical reality is what we seek to define when we try to go down from personal reality to make sense of the movement of the planets or the effects of drugs on the brain. And social reality is a way of talking about communities of persons with schemas in their heads which lead to certain patterns of agreement in overt behavior. But I don't see social reality as something external in the sense of historically determining what people do. The dynamic for me is at the level of how individual people interact with other people and with the material products of society, and how they change their schemas on that basis. In some cases those changes make them outcasts and in others those changes begin to cohere with other individuals' changes until finally we see a revolution or a whole change in the community or in social structure.

Chapter 1 The Embodiment of Mind

THROUGH THE CENTURIES, more and more of the properties of matter have been explained by physics and chemistry. Subtle new concepts like mass, the electromagnetic field, and the electron have first been postulated as theoretical entities and then have become accepted as part of physical reality. This happens because they become integral parts of theories of the physical world which are intellectually compelling and which meet the pragmatic criterion of successful prediction and control.

But what if we turn from the reality of physical matter in motion to the reality of the person? Can we still hold to the naturalist view that there is only spatiotemporal reality, or are there realities—like the mind or will or soul of the individual, or the historical sweep of social forces, or the one, true, God—which transcend space and time?

In this chapter, I introduce *cognitive science* as the vehicle for a naturalist account of the person. Yet the current state of cognitive science is no more (and perhaps less) complete an account of human cognition than was Kepler's a complete account of matter in motion. Thus, in succeeding chapters, we shall confront the present status of this science with the debates over free will, with Freudian myths of the ego, id and superego, and with the realities of language, society, ideology and religion to ask: to what extent can cognitive science give a theory of the person? And to what extent, if any, does the reality of the person transcend what cognitive science can hope to explain?

I take cognitive science to be a loose federation of work in artificial intelligence (AI), linguistics, psychology and neuroscience.[1] More specifically, I use it as an umbrella term which unites three areas: AI, "the attempt to program computers to do things that you would swear require intelligence until you know that a computer has been programmed to do them"; cognitive psychology, which uses the language of information processing to design models which can be run on computers to emulate the overt behavior of a human performing some intelligent task; and, at the finest grain of detail, brain theory, which must not only explain behavior patterns of animals and humans, but must incorporate data on brain function and neural circuitry. What these have in common is that they yield models of cognition and intelligence that can be run on the computer. Thus, to the extent that cognitive science succeeds, we will have theories of mind that are operative. The machine can *simulate* intelligence. This raises the question: can an artificial intelligence—a computer, whether programmed ad hoc, or using principles culled from behavioral or neurological observation—actually *exhibit* intelligence. To put this in perspective, note that it is an open question for a particular property as to whether or not a simulation itself exhibits that property. In the case of motion, a computer simulation of a moving arm will not exhibit motion, but a robot is a simulation that does exhibit the simulated property.

Human and Artificial Intelligence

My answer to the question, "Can machines exhibit intelligence?" and its cognate, "Can human intelligence be given a naturalistic explanation, without invoking a nonspatiotemporal mind or soul?" will be twofold. I will accept that AI is limited now, but will counter certain philosophical arguments that AI is limited in principle. Nonetheless, when we counterpose intelligence in abstracto with what is specifically human about the way in which we are intelligent, I shall argue that we must take into account the way in which we are embodied, both in having human bodies and in being participants in human society.

The philosopher John Searle (1982) has argued that an AI system could only emulate intelligence, not exhibit it. As his straw man, he took Roger Schank's theory of scripts (Schank and Abelson, 1977), an account of the way in which we can go beyond the information given when answering questions. We bring general knowledge to bear. Schank has formalized this in terms of "scripts" which incorporate knowledge of certain types of situations. For example, the "restaurant script" would let us infer that a waiter or waitress had served the food, even if this was not explicitly mentioned in a specific account of dining out. Schank would want to argue that a computer, equipped with enough scripts and control programs to process incoming stories to generate answers to questions, would exhibit intelligence. Searle counters this claim as follows. Imagine that, instead of a computer equipped with programs to take in stories and questions and to print out answers, we have a human who only understands English sitting in a large box. Imagine, further, that the input and output are in

Chinese and that the man has a set of rules which do not involve translation into the English he understands, but rather tell him how to arrange the, to him meaningless, Chinese characters to generate the output. The idea is that the rules would correspond, formally and exactly, to those that Schank gives the computer. Searle argues, I think convincingly, that the man in the box would in no way understand the story. Searle then concludes that Schank's program cannot, therefore, understand the story and thus does not exhibit intelligence.

At first sight, this argument against a machine exhibiting intelligence is compelling. But then Searle begins to answer his critics. Imagine that, instead of symbols being fed into a computer, there was a TV camera, so that pictures served as input. Imagine that instead of symbols as output, the computer could move effectors and was, in fact, a robot. Add a more and more complex data base to the system. Even replace the computer circuitry with an electronic analogue of a neural net. With each step, Searle argues that the expanded system can still only simulate intelligence, not exhibit it. I am reminded of a classic story about Norbert Wiener, the "father" of cybernetics (an earlier incarnation of cognitive science, cf. Wiener [1961], Arbib [1975]). Wiener believed he had resolved a famous nineteenth-century mathematical conjecture—the Riemann hypothesis—and mathematicians flocked from Harvard and MIT to see him present his proof. He soon filled blackboard after blackboard with Fourier series and Dirichlet integrals. But as time went by, he spent less and less time writing, and more and more time pacing up and down, puffing at a black cheroot, until finally he stopped and said, "It's no good, it's no good, I've proved too much. I've proved there are no prime numbers." And this is my reaction to Searle. "It's no good, it's no good, he's proved too much. He's proved that even *we* cannot exhibit intelligence."

Note that Searle is not denying that there is a naturalist explanation of the mind, but he does want to show that only a computer with the biochemical structure of the human brain can exhibit intelligence. To look at this from another perspective, recall the bald man paradox. "If a man has no hair on his head, he is certainly bald. But if a man is bald, adding one hair to his head will not remove his baldness. But from zero we can pass to any number by adding one sufficiently many times. Thus, no matter how many hairs a man has on his head, he must be bald." The conclusion must be that baldness is not a predicate with a hard cut-off at some exact number of hairs. A few hairs and there is baldness, many hairs and there is no baldness, but there is no sharp transition in between. Evolution gives us a similar view of intelligence: the amoeba is not intelligent, the human is, and we can draw evolutionary trees which carry us from one to the other, without there being any single evolutionary step that converts an unintelligent species into one that does exhibit intelligence. I believe we can make a similar claim for AI systems. At most, today's systems can exhibit but limited *aspects* of intelligence. But we have not seen any argument to block the continuing development of such systems until they exhibit intelligence in some very rich sense of the term.

Nonetheless, I do want to argue that there is something distinctive in the way in which we, as humans, are intelligent. We have already discussed the transition from a pure symbol-manipulation system to one that interacts with the world. A dictionary can tell you what a book is only in the limited sense of giving a definition in terms of other words, and if you do not know all these words, you will refer to further definitions in terms of yet other words. The dictionary never breaks out of the vast "web of words." But you come to understand the new word, to some extent at least, because enough of the other words you encounter make contact with your own lived experience. You come to understand the concept of book to the extent that you can look at one and recognize it. But this recognition is not limited to naming the object; you also know that you can pick the book up, turn the pages, read the writing, look at the pictures, and gain information or entertainment therefrom. We go beyond the formal structures, the semantic networks, of artificial intelligence to root our knowledge in interaction with the world.

The evolution of linguistics may be related. The modern era of linguistics was initiated by Chomsky's theory of syntactic structures,[2] of the formal relationships required between words to make them constitute a legal sentence of the language. Joan Bresnan (1978) was amongst those who built on Chomsky's insights by developing the concept of the lexicon, the idea that the general rules of grammatical structure are not enough to characterize the patterns of language without much information about the particular roles that specific words can play, including certain semantic roles, such as which words can stand in the role of "instrument" or "agent" for a particular verb. We might see much of the AI work on knowledge representation, including Schank's scripts, as an attempt to go from the lexicon to the encyclopedia, an almost open-ended compilation of the background knowledge that may implicitly shape our sentences and discourse. What we are adding here is that much of this knowledge goes beyond the formally codified. We speak of the embodied mind, enriched by the experiences of living within a human body as a member of a human society.

With this, let us recall Shylock's impassioned defense of *his* humanity:

> Hath not a Jew eyes? Hath not a Jew hands, organs, dimensions, senses, affections, passions? Fed with the same food, hurt with the same weapons, subject to the same diseases, healed by the same means, warmed and cooled by the same winter and summer, as a Christian is? If you prick us, do we not bleed? If you tickle us, do we not laugh?

> If you poison us, do we not die? And if you wrong us , shall we not revenge?
>
> —SHAKESPEARE, *The Merchant of Venice*, Act 3, Scene 1

This is the person reality in which our lives are rooted. It is his *experience* that makes Shylock human. What is amazing about this quotation is how much of that experience Shylock shares with animals, but he is distinguished by his conscious knowledge of much of this experience. We stress again that theories of the mind that treat it as disembodied falsify the nature of our cognitive lives.

In later chapters we shall go beyond this level of embodiment to take account of the fact, which in Shylock's speech is only expressed in his talk of revenge, that humans are social creatures, and we shall have to understand how our reality is expressed as a network of levels of reality, from the physical to the social.

Gödel's Theorem and the Role of Logic

In the early part of this century, philosophers and mathematicians pondered the question of whether there existed a formal, logical system in which, starting from axioms and using certain rules of inference, one could deduce as a theorem, i.e., as something formally provable within the system, each true statement (and no false statements) about arithmetic, everything that might be true about adding and multiplying numbers, and so on.[3] Gödel (1931) studied logical systems which were *adequate*, in the sense that they had enough expressive power to state the truths of arithmetic whether or not they could be proved, and *consistent* in the sense that it was never possible to prove both a statement and its negation within such a system. In 1931, Gödel shattered the expectations of many eminent men, from Hilbert to Russell, by proving that any arithmetical logic that was both adequate and consistent was *incomplete* in that there were true statements that could be expressed but could not be proved—neither the statement nor its negation could be proved within the system.

This was a result of great importance in the philosophy of mathematics, but some philosophers, e.g., Lucas (1961) have tried to apply it in the philosophy of mind, claiming that it shows that a machine cannot think. This is a bizarre approach, because it can only apply to a mind-model that consists of a machine with all its knowledge carefully coded in logical form at some particular time and for which the only "mental operations" would consist of making strict deductions from the information encoded in it from the beginning. However, all cognitive scientists would agree that an adequate model of the mind would represent it as open to novel experience. One of the basic tasks in modeling mental activity is to understand our ability to learn from our mistakes. But if we make mistakes, we have certainly transgressed the limits of consistency. Certainly, machines have been built that incorporate learning algorithms, and Gödel's theorem would seem to say nothing about their limitations.

Let us look at this in another way. Gödel's incompleteness theorem says that if you start with consistent axioms and apply the rules of inference, then the collection of theorems that can be deduced is incomplete in that not all true statements are provable. But there is some sense in which beings who live within the world are forced to be "complete," in that a decision as to which action is appropriate often cannot be postponed. If we are crossing the road and a car is bearing down upon us, we will not survive very long if we require a long period of careful theorem proving before deciding whether or not it is true that the best action in this circumstance is to stay where we are or to jump out of the way. Rather, as beings in the world, we must make many decisions with respect to certain real-time limitations. Thus there is a sense in which time limitations force some form of completeness upon those statements that we must articulate in our actions. Gödel's theorem may be taken to say that if your actions in some sense embody complete logic, then that logic must be inconsistent. I thus assert that the import of Gödel's theorem for formal models of the mind is simply that if you have to make decisions, you will make mistakes. This observation is not a sound basis on which to distinguish human from machine!

I have been somewhat satirical about a purely logical model of the mind, with mental life restricted to the inferring of conclusions from formal statements. But I do not want this attack to be seen as a rejection of logic. I do not see logic as the sine qua non of human thought, but certainly see it as an important limiting case. I want to suggest that humans do not live on the whole by rigorous arguments, but rather make decisions that seem plausible to them given what they happen to be. Consider an example of logic in everyday life. Here is a rule of inference in formal logic: if it is the case that the assertion "A or B" is true but we know that "not A" is true, then we may infer that "B" is true. Now imagine the following scenario: I always leave my sunglasses in the bedroom (A) or the kitchen (B). I look around the bedroom and I cannot find the sunglasses, so assertion "A" is false. I may thus deduce "B," that the sunglasses are in the kitchen. I go to the kitchen, search desperately, yet do not find these sunglasses. Are we to decide that the foundations of logic lie shattered? No, we conclude that logical rules of this kind, while useful in certain formal situations, provide only a first approximation to our everyday logic. The everyday rule is that I almost always leave my sunglasses in the kitchen or the bedroom; if I don't find them in the kitchen I should look in the bedroom and vice versa. But if they are not there, I then search my memory or environment more widely, and this "search" may not be fully captured by the strict application of logical rules. Even though that logical deduction is an important limiting case of our decision making, it is in some sense exceptional. Human decision making is embedded within a network of schemas, so that an analogical schema-based process of cooperative computation comes closer to characterizing mental behavior than does logic.

Schema Theory

Earlier, I spoke of cognitive science as having three components: the ABC of Artificial Intelligence, Brain Theory, and Cognitive Psychology. Yet it must be admitted that the majority of workers in cognitive science have little interest in brain or action, and that much of their work focuses on linking AI and cognition to symbol manipulation in general and to linguistics in particular. My own work, on the contrary, tries to see our linguistic abilities as rooted in our more basic capabilities to perceive and interact with the world (Arbib, 1972). I use the term "schema theory" to designate this approach to cognitive science, for it uses the term "schema" to denote the basic functional unit of action and perception (Arbib, 1981a). What, then, is this schema theory in which we are to give an account of the embodied mind, an account which is to transcend mind/body dualism by integrating an account of our mental representations with an account of the way in which we interact with the world?

The history of schemas goes back to Immanuel Kant and beyond, but I want to start with the work, at the beginning of this century, of the neurologist Sir Henry Head (one of those fortunate people whose names signal their professions). Head and Holmes (1911) discussed the notion of the body schema. A person with damage to one parietal lobe of the brain may lose all sense of the opposite side of his body, not only ignoring painful stimuli but neglecting to dress that half of the body; conversely, a person with an amputated limb but with the corresponding part of the brain intact, may experience a wide range of sensation from the "phantom limb." Even at this most basic level of our personal reality—our knowledge of the structure of our external body—our brain is responsible for constructing that reality for us. Our growing scientific understanding of knowledge takes us far from what "common sense" will tell us is obvious. One of Head's students was Frederick Bartlett who, in 1932, published *Remembering*. Bartlett noted that people's retelling of a story is not based on word-by-word recollection, but rather on remembering the story in terms of their own internal schemas, and then finding words in which to express this collection of schemas.

Such ideas prepare us for the work of Kenneth Craik who, in *The Nature of Explanation* (1943), views the nature of the brain to be to "model" the world, so that when you recognize something, you "see" in it things that will guide your interaction with it. There is no claim of infallibility, no claim that the interactions will always proceed as expected. But the point is that you recognize things not as a linguistic animal, merely to name them, but as an embodied animal. I will use the term "schema" for the building blocks of these models that guide our interactions with the world about us. To the extent that our expectations are false, our schemas can change, we learn. We then see such writers as Richard Gregory (1969), Donald MacKay (1966) and Marvin Minsky (1975) building upon this notion of an internal model of the world, at first in the cybernetic tradition, to develop the concept of representation so central to work in AI today.

One of the best-known users of the term "schema" is Jean Piaget, the Swiss developmental psychologist and genetic epistemologist.[4] He traces the cognitive development of the child, starting from basic schemas that guide his motoric interactions with the world, through stages of increasing abstraction that lead to language and logic, to abstract thought. Piaget talks both of *assimilation*, the ability to make sense of a situation in terms of the current stocks of schemas, and of *accommodation*, the way in which the stock of schemas may change over time as the expectations based on assimilation to current schemas are not met. These processes within the individual are reminiscent of the way in which a scientific community is guided by the *pragmatic criterion* of successful prediction and control (Hesse, 1980). We keep updating our scientific theories as we try to extend the range of phenomena they can help us understand. It is worth noting, however, that the increasing range of successful prediction may be accompanied by revolutions in ontology, in our understanding of what is real, as when we shift from the inherently deterministic reality of Newtonian mechanics to the inherently probabilistic reality of quantum mechanics.

There are a number of studies in the literature of cybernetics and artificial intelligence that use such names as frames, schemas, and scripts to describe the processes whereby knowledge is represented. In my own group, we have used schemas to provide a functional level of analysis of what goes on in the brain of an animal during sensorimotor coordination; to provide intermediate-level programs for mediating between vision and touch and the control of the movements of a robot; and (as we shall see in chapter 2) we have used schemas in formal models of language acquisition and production. It may be somewhat presumptuous, then, to talk as if there is a body of science that answers to the name of schema *theory* if the word "theory" is taken to imply that there is some agreed upon core of time-tested definitions and theorems. Rather, as the above examples suggest, schema theory (like the cognitive science of which it is a part) is a loose federation of "mini-theories," each claiming some success, but with tensions between them that provide many challenges for future research. We can look at different examples of schemas and suggest that there is a richness that grows as more and more scientific contributions are made.

It may be useful to compare our use of the word "schema" with the way in which the word "program" is used in computer science. We can recognize a particular program for adding up a column of numbers or for inverting a list. But if we try to define a program in complete generality, we are in trouble. Do we mean a program in FORTRAN, a program in PASCAL, a program in LISP? Do we mean a serial program or a concurrent program? At any particular time, computer science has no single abstract definition of a program. Nonetheless, we recognize commonalities in the current set of program types so that as we develop new ideas of programs we can build upon the knowledge already attained. That would be my claim for a theory of schemas as well. More specifically, I would assert that schemas are programs developed to satisfy the following criteria (Arbib, 1981a):

> (1) They serve both to represent perceptual structures and to subserve distributed motor control.
>
> (2) Schemas may be instantiated. We may have a schema that represents our generic knowledge of a single chair. We need several *instances* of that schema, each suitably tuned, to subserve our perception of several chairs.
>
> (3) The programs are concurrent. Unlike the programs of most present computers that carry out a series of instructions one after another, we postulate that the brain can support the simultaneous concurrent activity of many schemas for the recognition of different objects, and the planning and control of different activities. ("Just because the brain looks like a bowl of porridge doesn't mean it's a cereal computer"!)

The term "schema theory" does not refer, then, to one polished and widely accepted formalism with the above properties, but rather covers a number of attempts by my colleagues and students to work within this general perspective. For work within artificial intelligence, including our work in machine vision and robotics, we ask how to define schemas as program units with the above properties that meet criteria for ease of implementation or for computational efficiency. For work within brain theory and cognitive psychology, schemas are designed to serve as units of complexity intermediate between behavior and neuron, and which help us "decompose" overall behavior in a fashion that gives us insight into the data of psychology and neuroscience. Here are a few examples.

My late friend Rolando Lara and I have developed schema theory in our study of neural mechanisms of visuomotor coordination,[5] paying special attention to behavioral, neuroanatomical and neurophysiological data on frogs and toads. We have looked at prey-predator pattern recognition, approach-and-avoidance behavior, and so on, and said "Let's try for an intermediate level of analysis of the frog's behavior in terms of interacting schemas and then ask how each of these schemas could be implemented in neural terms." This is a contribution to that part of schema theory that talks about input-matching (pattern recognition) routines, motor routines, and their competition and cooperation.

In studying the way in which the toad detours around a barrier to reach a worm, we look at the interaction of perceptual schemas for prey-recognition and barrier-recognition, the formation of depth maps, and the control of motor schemas for such unit behaviors as sidestepping, orienting, and snapping. In using schemas to explain the behavior of animals, we are close to the work of ethologists like Lorenz and Tinbergen who talked about animal behavior in terms of innate releasing mechanisms. What we are doing here is twofold: firstly, we are trying to put schema interaction on a firm computational basis, and secondly, having done that, we are trying to see how brain mechanisms, neural networks, could explain what we have just described in computational terms.[6]

Another of our approaches to schema theory comes with the task of programming hand movements for a robot. What schemas does it need? What sensory information does it have to have? How is it going to transform it? I reach out for something, and that same movement of reaching out is both moving my wrist to carry the hand but is also preshaping the hand. So here again we talk about concurrent activity of schemas. When will one schema say it has completed its job and activate another to take over? When will two schemas be activated at the same time? In this case we could say there are two schemas, one for the movement of the hand, and one for the shaping of the hand. Those two schemas are activated at the same time, but to succeed they have to share some information. The schema for moving the wrist to carry the hand has to know where the object is, while the schema for shaping the hand has to know the size and orientation of the object. Proceeding further, we can analyze these subschemas in terms of yet smaller schemas which direct basic grasps, structuring them in terms of oppositions between "virtual fingers," groups of fingers moving together as a unit within that specific portion of the task.[7]

In another study, with Jane Hill (Hill and Arbib, 1984), we have done more Piagetian work. We looked at how a two-year-old child acquires language and gave a theory of how the child's schemas can represent its ability to use language. We saw how the child assimilates what it hears to produce its response, and yet in the process accommodates, schemas changing so that the child's language matures. We have been able to provide a causal explanation which is consistent with the general pattern of Piaget's findings, but which gives us more insight into underlying mechanisms. We shall have more to say of this and other models in chapter 2.[8]

With these examples before us, we can begin to see how schema theory is developing to give us a cognitive science that can address issues in human perception and movement, and in language and learning. The work of my group on schema theory is silent about the world of feelings and emotions, and we shall say more of this when we discuss Freud in chapter 3. For now, I want to concentrate on perception and movement. A key concept is that of the *action/perception cycle* (Arbib, 1981a; Neisser, 1976), viewing humans not as stimulus-response creatures who wait passively until, hit with a stimulus, they emit some corresponding reflex response. Rather, there is always a *schema assemblage* of currently active schemas, built from the network of schemas that represent our "knowledge," that constitute our representation of our current goals and situation. This schema assemblage guides our actions, but is in turn updated as our actions lead to new perceptions; and the interaction of events and expectations drives the process of accommodation whereby our total network of schemas is updated and expanded.

To make this more concrete, let us consider VISIONS, the machine vision project led by my colleagues Allen Hanson and Edward Riseman. The input to the computer encodes a color photograph of an outdoor scene of trees, houses, grass, sky, and so on. How do we get the computer to recognize these elements in the scene and to correctly label the parts of the image to which they correspond? The first step is called *segmentation*, breaking the scene into different segments or regions which may serve as candidates for meaningful parts of the picture. Cues for segmentation might include discontinuities of color or texture or (with stereo pairs) depth; while commonalities of color or texture might suggest picture elements to be aggregated into a single region. Unfortunately, even the best of such procedures of "low-level" vision cannot yield a perfect segmentation: perhaps a shadow on a wall caused it to be segmented into several regions;

maybe the highlighting on a roof of bluish gray slate led that roof not to be adequately segmented off from the sky. At this stage, the computer must use programs for "high-level vision" that can invoke *perceptual schemas* which embody knowledge about the objects that may occur in the scene (Hanson, Riseman, Weymouth and Griffith, 1985). The bluish region at the top of the picture *may* be sky; a few contiguous regions that can be aggregated into a rectangle *may* be a good bet for a window or a shutter or a door; a portion of a parallelogram *might* be the boundary of the roof; and if there is a good bet for sky just above it, or a good bet for shutters just below it, then the level of confidence in the roof-hypothesis increases.

The machine vision program is not an all-or-none process. Some regions may be provisionally joined together, some may be split apart. Different interpretations will have different confidence levels. There will be a process of *cooperative computation* between multiple hypotheses, strengthening some and weakening others, to converge upon a coherent interpretation of the whole image. My claim is that similar processes occur in our brains as multiple brain regions—each composed of millions of concurrently active neurons—interact to commit us to a single overall course of action (Arbib, 1976, to appear b; Arbib, Overton, and Lawton, 1984).

I might suggest that to make sense of any given situation we call upon hundreds of schemas in our current "schema assemblage" and that our lifetime of experience, our skills, our general knowledge, our recollection of specific episodes, might be encoded in a personal "encyclopedia" of hundreds of thousands of schemas, enriched from the verbal domain to incorporate the representations of action and perception, of motive and emotion, of personal and social interactions, of the embodied self. It is in terms of these hundreds of thousands of schemas that I would offer a naturalistic account of the self, embodied in space and time. Nonetheless, for many people this raises the question: could hundreds of thousands of schemas, or billions of neurons, cohere to constitute a single personality, a self, a personal consciousness?

Reductionism

I will try to approach this question through a critique of reductionism, the attempt to reduce the laws of one science to the laws of another, as we might explain parts of chemistry in terms of the laws of physics. I want to start by looking at *one-way* reductionism, which, in our present case, would be the claim that all the laws of the mental are deducible from the laws of the physical brain; and that there is a basic science, call it neurophysiology, from which all of psychology could be deduced, if only we did sufficient computation. To approach this, let us ask whether such one-way reduction is possible even within physics.

Consider the example of statistical mechanics which explains bulk properties of matter by appropriate averaging over the interactions of myriad constituent particles. To conduct this process of explanation, we must be guided by phenomena observed at the macrolevel. Some phenomena, as in the history of the photoelectric effect or spectroscopy, defy explanation in terms of current microphysics, and so demand the redefinition of underlying physical theory (Pais, 1982). In other cases, as in the modern study of magnetism, we are not forced to new microtheories, but do have to make new analyses of statistical properties and fluctuations to see how the observed phenomena can be deduced from the underlying theory. And this may yield paradoxes, as when averaging takes us from a reversible Newtonian mechanics to an irreversible thermodynamics. What seemed at first a purely deductive process from micro to macro seems once again to have changed our ontology. The result of all these studies—both new underlying assumptions and new methods of deduction and approximation that propagate "upward"—allows us to construct new realities in the macroworld, such as computer circuitry or atom bombs.

In summary, then, we see that no set of laws has arisen at one level which serves as the ultimate arbiter of what constitutes reality at some other level of observation and control. This leads us to the more adequate concept of *two-way reductionism* in which the two sciences of (statistical) mechanics and thermodynamics are modified by the attempts at reduction to yield a new science which extends, but is not identical with either of, its prereductive components.

This two-way reductionism is the way in which I would regard cognitive science. We do not assert that we already have a complete theory of neurons or schemas to which, if only we could compute enough, we could reduce all that is to be known about human cognition and personal reality. Rather, we view schema theory and cognitive science as evolving in response to critiques of the limitations of our current understanding of mind and person.

With this, let us briefly examine the diverse roles that schemas will have to play if we are to build a truly satisfactory theory of how persons come to know their reality.

We have come to think of a schema as a unit for the construction of our representations of reality, but not necessarily as an atomic unit. In the same way, a computer program may be used as a subroutine, a unit in building larger programs, and may itself have been built up from other programs in its turn. What we add to the notion of a program for a serial computer is that of concurrency, of many schemas being active at the same time; of the embodied subject who requires schemas for action and perception; and our stress that schemas constitute a network that brings together our notions of reality at many different levels. We have the neurophysiological level, where the brain theorist seeks to instantiate schemas in terms of neural networks; and the cognitive psychologist's analysis, in information-processing terms, of schemas for basic pattern recognition or memory tasks.

To move from this level of analysis to an understanding of the human individual, we will need to study the coherence and conflicts within a schema network that constitutes a personality, with all its contradictions, as when we look at Freud's concept of identification as providing person-schemas; and the holistic nets of social reality, of custom, language and religion. It is interesting that, quite independently, Marvin Minsky (1975) chose the word "frame" for his AI account of schemas for such social situations as a birthday party while Erving Goffman (1974) chose the title *Frame Analysis* for his sociological analysis of the way in which people's behavior depends on the frame, or social context, in which they find themselves. One example. A patient enters the doctor's office, and the doctor asks "How are you?" and the patient replies "Fine, thanks." After they are both seated, the doctor again asks "How are you?" and now the patient replies "Doctor, I have this terrible pain. . . ." What changed? The action moved from the "greetings frame" to the "doctor-patient frame."

The following points summarize and extend our understanding of schema theory as an open-ended subject, responding to, but changing, our concepts of the reality of our personal and social worlds.

(1) There is an everyday reality of persons and things. If you cut a person, he or she bleeds. If you drop a kettle, boiling water may scald you. Love can turn to jealousy. How can we come to know this reality?

(2) Schema theory answers that our minds comprise a richly interconnected network of schemas. An assemblage of some of these

schemas represents our current situation; planning then yields a coordinated control program of motor schemas which guide our actions. As we act, we perceive; as we perceive, so we act.

(3) Perception is not passive, like a photograph. Rather it is active, as our current schemas determine what we take from the environment. If we have perceived someone as a friend, we may perceive their remark as a pleasing joke; yet the same words uttered by someone we dislike, even if she too intended a joke, may be perceived as an insult that elicits a vicious response.

(4) A schema—as a unit of interaction with, or representation of, the world—is partial and approximate. It provides us not only with abilities for recognition and guides to action, but also with expectations about what will happen. These may be wrong. We sometimes learn from our mistakes. Our schemas, and their connections within the schema network, change. Piaget gives us some insight into these processes of schema change with his talk of assimilation and accommodation.

(5) There is no single set of schemas imposed upon all persons in a uniform fashion. Even young children have distinct personalities. Each of us has very different life experiences on the basis of which our schemas change over time. Each of us thus has our knowledge embodied within a different schema network. Thus, each of us has constructed a different world-view which each of us takes for reality. This observation will be very important when, in later chapters, we try to reconcile the schemas of individual and society.

With this, we have come to understand the diversity of possible schemas. We have seen that they may rest on individual style, yet, as we shall shortly see, they can be shaped by the social milieu. A network of schemas—be it an individual personality, a scientific paradigm, an ideology, or a religious symbol system—can itself constitute a schema at a higher level. Such a great schema can certainly be analyzed in terms of its constituent schemas, but—and this is the crucial point—once we have the overall network, these constituents can find their full meaning only in terms of this network of which they are a part.

This hierarchical view of schemas is very close to the views of C. S. Peirce on what he calls "habits" (cf. Burks, 1980). For Peirce, a habit was any set of operative rules embodied in a system. He emphasized—and this anticipates Piaget—that they possess both stability *and* adaptability. He had in mind an evolutionary metaphor: species form a stable unit for our analysis of the present state of the animal world, and yet we know that these units are subject to evolutionary change. Thus Peirce's habits, like our schemas, can serve as building blocks in a hierarchy of personal rules, of society, science and evolution, and yet may themselves change over time.

In this way, we view knowledge as inseparable from an evolving schema network: there is then no such thing as *sure* knowledge, and we must reject the philosopher's definition of "knowledge" as "true belief." Certainly, there are schemas that embody what we *now* take to be true. But schemas also include representations that the naturalist would not want to call knowledge. There are false, but useful, models of the natural world; ideal types of human behavior and social order that may never be realized, including utopias, heavens or hells; and there are ideals as given by schemas for the beautiful, the true and the good, which are very real as determinants of the way in which we behave, whether or not we believe they represent some external reality.

The question remains: Even if we accept the claim that schemas provide the right elements with which to build a naturalist theory of minds, will it be possible for an extended schema theory to give an adequate account of person-reality without invoking transcendent categories? There are certainly schemas of God, but is there a God of which these schemas are but a pale spatiotemporal caricature? Or is reality purely in space and time, with such schemas providing myths and fictions that may prove more or less useful in regulating our personal and social lives? To better prepare us for the answers that the remaining chapters may provide, let me summarize our current understanding as follows:

In physics, we have seen continual improvement according to the pragmatic criterion of successful prediction and control. But though quantum theory and the theory of relativity preserve the pragmatic successes of Newtonian physics, they have changed our view of the underlying reality. Yet, despite these ontological revolutions in science, we find that apples continue to fall and the planets continue in their paths. Thus, even though we know that the theories of physics, like our individual schemas, are open to change, most of us would agree that there is an external *spatiotemporal* reality independent of human constructs, which provides the touchstone for our attempts to build physical theories.

On the other hand, we can see that many of the patterns of our daily lives—whether our own idiosyncracies or the conventions of a language community or a social group—are purely human constructs. For example, in Massachusetts we must drive on the right-hand side of the road; in Australia on the left. But this is a purely social convention: life is easier if the members of a single driving community observe the same basic rules. We do not hold that there is a greater Reality, the one true Road-Reality, which the framers of human road rules try to approximate.

What then of human reality? Is there a God-reality that transcends space and time, and in which alone human life can find meaning? Are there rules of social development and historical process which inevitably shape human nature? Is our reality to be found in the free will of the individual, and is this essence of the human unconfined by the shackles of physical law? Or, despite all these alternatives, will our growing understanding of the human reshape again our concepts of spatiotemporal reality while letting us come to understand humans as biological mechanisms that must wrest their own meaning from the historical contingencies of biological and social evolution?

In subsequent chapters, we must expand our schema theory to meet the challenge of the human will, of language, of ideology and religion before we can properly address these questions. In any case, we have already come far enough to see that our schema theory will not reduce people to indistinguishable computer programs. The uniqueness of endowment and experience yields the unique schema network which makes each person an individual whose reality is only in part socially constructed. As individuals within society, we may understand authority, but we need not invariably accept it. And to that extent, we are free.

Chapter 2 Language: Individual and Social Schemas

WE FIND, at the core of our concern, the ineradicable tension between language and the richness it seeks to express. As Shattuck (1984) observes in his review of Nathalie Sarraute's book *Childhood*, Sarraute says "There is always a kind of drying out produced by language. . . . For me . . . there is something prior to language: a sensation, a perception, something in search of its language, which cannot exist without language." What are we to make of that last phrase, "which cannot exist without language?" For she sees language, not only as expressive but also as destructive: "Scarcely does this formless [feeling], all timid, and trembling try to show its face than all powerful language, always ready to intervene so as to re-establish order—its own order—jumps on it and crushes it." The problem is that the writer, and of course the speaker, must try to express within words that which goes beyond words. When Sarraute recalls a particular childhood moment, she cannot accept the terms "happiness" or "ecstasy." Even the simple word "joy cannot gather up what fills me, brims over in me, disperses, dissolves, melts into the pink bricks, the blossom-covered espaliers, the lawn, the pink and white petals, the air vibrating with barely perceptible tremors, with waves . . . waves of life, quite simply of life, what other word? . . . of life in its pure state. . . ." (Sarraute, 1984: 57)

The previous chapter sketched how networks of schemas could integrate the manifold experiences of the embodied self, offering a schema theory that is somehow to provide a scientific approach to the richness that words fail to express. But, of course, any formalism, and that includes schema theory, that separates a part of the network from its connections inside and outside must itself do damage to the plenitude of lived experience. Does this damn our efforts at building a science of the mind? I think not, as may be seen by recalling a fiction of Jose Luis Borges, called "Of Exactitude in Science." Borges (1975) talked of a country that prided itself on its cartographical institute and the excellence of its maps. As the years went by, this institute would draw maps of greater and greater accuracy until at last the institute achieved the ultimate, the full scale map. And, Borges says, if you wander through the desert today, you can see places where portions of the map are still pegged to the region they represent! The point of all this, of course, is that our job as cognitive scientists is only to *chart* the territory of mental life to establish the major phenomena and their relationships; not to provide the full-scale map, not to replace a life richly lived by the running of some computer program. Nonetheless, we have offered schema theory as the grounding for our epistemology and our task in the present chapter is to begin to extend it to the analysis of such social phenomena as language, ideology, and religion. In doing this, it is appropriate to chart a tension within schema theory that echoes that charted by Nathalie Sarraute—tension between the current formal models of limited phenomena (whether expressed in computer terms, mathematical equations, or neural networks) and the richer description of human experience that we have in our everyday language. Our job as scientists, irrespective of our job as philosophers, is twofold: not only to provide explicit accounts where we can, but also to understand the limitations of those accounts. And so we always exist in that tension, that dialogue, between the charted and the unknown. In sensing that tension, some may argue that the current limitations will in due time be removed by further scientific research, while others may argue that the limitations are not temporary but are irremovable in principle. In chapter 1, for example, we discussed Searle's critique of the strong AI claims of Roger Schank and criticized the use that some people have made of Gödel's incompleteness theorem as an argument for the shortcomings of the artificially intelligent. But here we want to see to what extent the somewhat individualistic account of schemas of chapter 1 can be extended to address social realities.

The Individual and Social in Relationship

In our earlier discussion, we have made explicit the tension between language and that richness of which even schema theory itself is a partial representation. Till now, we have stressed the individual level; language has been something that an individual can use, possibly to communicate with another individual. The time has now come for a key transition, from schema theory as a description of the mind of the individual to a schema theory that addresses the apparent reality of social forces and institutions.

When the South African novelist J. M. Coetzee was a graduate student at the University of Texas, he was staggered to discover "that every one of the 700 tongues of Borneo was as coherent and complex and intractable to analysis as English" (Coetzee, 1984). He found this sudden appreciation of the intrinsic value of each such language in and of itself to be an "odd position for someone with literary ambitions, . . . ambitions to speak one day somehow in his own voice," odd "to discover himself suspecting that languages spoke people or at the very least spoke through them." It is this very oddness that is at the heart of this chapter: the paradox of the individual actor discovering that much of what he took to be his individuality appears to be the playing out of social structures.

Our task, then, is to develop schema theory so that it may serve as a tool for resolving this paradox. The problem can, perhaps, be expressed in a different way: "How does one become a member of a particular community?" whether it be a language community, or a social community, or a religious community. In our terms, "How does the individual acquire the schemas that constitute, or construct, his social reality?"

An Individualistic Account of Schema Acquisition

To proceed, I must distinguish between a schema as an *internal* structure or process (whether it is a computer program, a neural network, or a set of information-processing relationships within the head of the animal, robot or human) and a schema as an *external* pattern of overt behavior that we can see when we look at someone "from the outside." When we read Piaget, we find ambiguity as he switches without warning from talking about the schema as if it were a structure inside the head which explains the behavior of the child to talking of the schema as something that the psychologist can observe.

This distinction, between the schema as internal and external, can be more formally understood by looking at the theory of finite automata, a branch of theoretical computer science. Very simply, a finite automaton is just a machine that receives, one at a time, symbols from some fixed set of inputs and produces, one at a time, symbols from some fixed set of outputs. What makes the theory of automata interesting is that these are not just stimulus-response automata. The present input symbol alone does not determine what the present output will be; a particular stimulus does not now elicit a fixed corresponding response. Rather, the automaton exhibits *sequences* of behavior, sequences of inputs corresponding to sequences of outputs. The *external* description of such an automaton characterizes how it behaves by specifying for each sequence of inputs the corresponding sequence of outputs, given that it starts in some standard state. This would correspond to the psychologist looking at the child and deciding that the next sequence of activity constitutes an example of a particular schema (as externally observed), a grasping schema or a suckling schema or the schema for object permanence. The *internal* description of the finite automaton explains its behavior in terms of a finite set of states, together with a specification of what happens when each input arrives, namely how the automaton changes state and, in changing state, which output it emits. The reason that we get different immediate responses to a given input is that different histories of input can drive the automaton to drastically different states.[1]

The task of the cognitive scientist, then, with schemas at the level of the individual is to infer, from externally observed regularities, internal structures that can provide an internal state explanation of them. But, of course, this raises problems. Within automata theory itself, there was the problem of knowing that the observed sequences of behavior were each initiated when the machine was in some standard state. But if you are observing people from the outside, how do you decide that they are in comparable states when you look at their behavior on different occasions? There is also the problem of deciding which pieces of behavior correspond to one schema and which correspond to another. We also note the problem, which we do not have in finite automata theory, of deciding what the input and output alphabets are, i.e., what the units are into which behavior is to be decomposed. In a recent novel by Anthony Burgess, there is one character who for many years thought that Elgar's "Pomp and Circumstance" was "Pompous Circus Dance"!

Those problems, however, are not just problems that the cognitive scientist faces in observing and rationally analyzing behavior; they are also problems that the child faces, perhaps unconsciously, when it comes to interact with its world. The child is in some sense trying to find out how to behave in such a way that it will achieve what it wants, that it will not get punished, and that it will gain some pleasure from its interactions with the world. We suggest that the child is trying to go from observed patterns of behaviors to internal patterns which provide appropriate representations. In chapter 3 we will discuss Freud's notion of identification, which we may see as the child's process of extracting from the parents' behavior schemas for behaving like the parent. To some extent this helps the child's growth, yet it also causes problems when the child incorporates into its schemas behavior which the parents abrogate for themselves, punishing the child for exhibiting them. This leads inevitably to certain tensions in schema acquisition and in mental development.

We now examine further Piaget's idea that the child has certain basic schemas and basic ways of assimilating knowledge to schemas, and that the child will find at times a discrepancy between what it experiences and what it needs or anticipates. On this basis, its schemas will change, accommodation will take place. It is, of course, an active research question as to what constitutes the initial stock of schemas. Much of Piaget's writing emphasizes the initial primacy of sensorimotor schemas, where other scientists, like Colwyn Trevarthen,[2] study the interactions between mother and child to stress social and interpersonal schemas as part of the basic repertoire on which the child builds. In either case, we can see that right from the beginning, the child has schemas—whether in terms of personal relationships or in terms of hunger and comfort—in which the, perhaps unconscious, knowledge of how to do something is inextricably intertwined with the knowledge of what to do. To the extent that we have skills to do something, we make an implicit value judgment that this something is worth doing.

Another important concept in Piaget's work is that of *reflective abstraction*. Piaget emphasizes that we do not respond to unanalyzed patterns of stimulation from the world. Rather, they are analyzed in terms of our current stock of schemas. It is the interaction between the stimulation—which provides variety and the unexpected—and the schemas already in place that provides patterns from which we can then begin to extract new operational relationships. These relationships can now be reflected into new schemas which form, as it were, a new plane of thought. And then—and this is the crucial point—since schemas form a network, these new operations not only abstract from what has gone before, but now provide an environment in which old schemas can become restructured. To the extent that we can form a general concept of an object, our earlier knowledge of a dog and a ball, and so on, becomes enriched. However, Piaget may be too sanguine in his view of the child progressing through stage after stage of increasingly coherent abstraction, which make sense of a richer and richer body of experience.[3] As we saw in our mention of Freudian identification, new schemas need not always just generalize or enrich old schemas. Again, we are reminded of our discussion of Gödel's theorem: the network is not a collection of consistent statements. Even though part of schema acquisition may reduce inconsistencies and seek out greater generality, inconsistencies may remain.

With all this, we conclude that the shared schemas within the individuals of a community may provide patterns of behavior that can provide regularities that allow the child to build schemas which will internalize for that child the patterns of the community. We continue to contrast the schemas outside, the observable patterns of behaviors, with these inside, the schemas as processes within the individual's head.

With this background, we can perhaps better appreciate the distinction between the schemas of the society, which the individual "holds" in his own head, schemas which embody his knowledge of his relations with and within society; and what I might call a *social schema*, a schema which is held by the society en masse, and which is in some sense an external reality for the individual. The following section on language acquisition is designed to help us understand what constitutes a social schema, in distinction to a schema within any one person's head, and how such a schema can affect what an individual does. As individuals come to assimilate communal patterns, they will provide part of the coherent context for others. But, the commonality of social behaviors still leaves space for discordances between individual and community, as we shall see in later chapters when we discuss Freud's *Civilization and Its Discontents* and when we look not at language but at ideology as a social structure.

Acquiring a Language

To be competent in English is, I would take it, not to interiorize some formal grammar, but to have the ability to communicate with others who speak English. The concept of English is an example of Wittgenstein's (1958) notion of "family resemblance." There is a definition of "dog" that I was told by my high school English teacher was in Johnson's dictionary but unfortunately doesn't seem to be there: "A dog is an animal which is recognized by other dogs to be such." This probably yields the best definition of an English speaker: someone who speaks a language intelligible as such to enough people who agree that they speak English. How does the child extract patterns from the utterances of a community of mutually intelligible speakers to form the schemas which interiorize his own version of the language? (I leave aside the fascinating study of what happens when the child grows in a bilingual setting and to what extent the child infers one composite language or discovers that there are two separate languages.)

A Chomskian approach to language acquisition would suggest that the child's innate schemas incorporate a great deal of the constraints of a transformational grammar, including such general categories as noun and verb, and certain constraints on the relationships between these categories.[4] I want to present an alternative approach (Hill, 1983; Hill and Arbib, 1984). This model of language acquisition in a two-year-old child starts, not from innate syntactic constraints, but from the observation that the child wants to communicate and likes to repeat sentences. However, and this accords well with our schema-theoretic basis, when the child "repeats" a sentence, it does not repeat the sentence word-for-word. Nor does the child omit words at random. Rather, the child's behavior is consistent with the suggestion that the child already has some schemas in its head and that an active schema-based process is involved in assimilating the input sentences and generating the simplified repetitions (recall the observations of Bartlett [1932]). To focus her study, Hill looked at a two-year-old responding to adult sentences either with a simple paraphrase or with a simple response. The child was studied once a week for nine weeks to provide a specific data base to balance the general findings in the literature. Intriguingly, the child changed every week. There was no such thing as "two-year-old language" to be given one lumped model. Since the child was different every week, the model had to be one of microchanges, in the sense that every sentence could possibly change the child's internal structures.

Hill's model provides a certain set of innate mechanisms that could drive the child's acquisition of a certain body of language. However, these mechanisms do not explain how it is that language eventually becomes nested or recursive in the sense that certain language structure can repeatedly incorporate simpler forms of that structure in increasingly elaborate constructions. Hill has outlined what those mechanisms might be, but has not studied them in depth. It does not appear that the elaboration of the model would force her to build in the structures that Chomsky would claim to be innate. However, it would also be premature to claim that her model is already sufficiently strong to invalidate Chomsky's claims about innateness. It *is* sufficiently strong to state that it is equally premature to accept those Chomskian claims.

At birth the child already has many complex neural networks in place and so in particular is able to suckle, to grasp, to breath, to excrete, to feel pain and discomfort, and so is able to learn that to continue a certain action in some circumstances and to discontinue another action in others is pleasurable. Note, however, that to say a schema is innate is not to imply that the adult necessarily has that schema. Once we begin to acquire new schemas they change the information environment of old schemas so that they can change in turn. We are not arguing here over whether language is or is not innate. We know that certain portions of the brain have to be intact for a person to have language and that language is degraded in specific ways by removing certain portions of the brain. What is at issue is to determine what it is that the initial structure of the brain gives to the child. Does it give it the concept of noun and verb, does it give it certain universals concerning transformational rules, or does it give, rather, the ability to abstract sound patterns, to associate sound patterns with other types of visual stimulation or patterns of action? Hill's model suggests that, at least for certain limited portions of a child's linguistic development, innate patterns of schema change can yield an increasing richness of language without building upon language universals.

In any case, let me briefly survey the "internal schemas" of the model: There were basic schemas for words, basic schemas for concepts, and basic templates which provided a grammar marked by a richness of simple patterns the child had already broken out of experience, rather than the grand general rules that we would find in the grammarian's description of adult language. And what was built in were not grammatical rules but rather processes whereby the child could form classes, try to match incoming words to existing templates, using those templates to generate the response.

Let me give one brief example of the way in which studies of this kind can give us insights. For a while, the child only produces two-word utterances. One simple way of accounting for this might be to have some notion of limited complexity which increases as the child matures. With such a model, one might next expect to see three-word utterances but, in fact, what comes next, though only for a week, is a predominance of four-word utterances which seem to be concatenations of two-word utterances such as "second ball, green ball." By the next week, instead of saying "second ball, green ball," the child was saying "second green ball." Hill's model explains this by invoking a process that collapses four-word utterances down to three-word utterances by deleting the first occurrence of a repeated word. This hypothesis would explain the earlier findings of Ed Matthei

(1979). (In fact, the example "second ball, green ball" was not something that Jane Hill's subject had said, but was chosen with malice aforethought from Matthei's study.) To probe the child's understanding, Matthei would place in front of the child a row of red and green balls, and ask the child to pick out the second green ball. The young child looks to see if the second ball is green, and is frustrated if it isn't! It may even rearrange the balls so that the second ball is green. It will pick up the second ball if it is also a green ball. This seems to accord with Hill's hypothesis that for the young child the semantics of "second green ball" is really given by the "flat" concatenation of "second ball" and "green ball," rather than by the "hierarchical" qualification of "green ball" by "second."

A Model of Language Production

As a second example of the tension between a particular formal description and a richer, but informal, description of some cognitive phenomenon, I want to look at a model of a fragment of language production (Conklin, 1983). The problem was to provide a computer with the ability to describe a picture of a scene in much the way a human would. This is no mere scanning of the picture from left to right and top to bottom, describing every possible object and relation encountered. Rather, humans seem to describe what is *salient*, and the most salient things tend to be described first. However, to simply state "there is a person," "there is a house," "there is a fence," and so on, in order of salience, is still not a very good model of a human's language production. Suppose there was a red door and a red car in the scene. Then, even though the car might come well before the door in terms of salience, a human speaker would recognize the commonality and so might produce a sentence that would combine the two items: "The house has a red door, and just down the road there's a red car as well." Conklin's description generator took as input a network expressing the names of the main objects of a scene and their relationships; each node or arc was labeled with a number that expressed the salience of the corresponding object or relationship. The system would then use both the salience ordering and rhetorical rules to extract a few salient objects and relations from the network at a time. A system called MUMBLE (McDonald, 1983) would than package them into a rhetorically well-structured sentence, with proper use of pronouns, and so on.

As we study a system of this kind, we see that certain aspects of human experience are captured in a way that is rather formal and yet does begin to show some of the richness of interrelationships between what we try to express and the way in which we say it. Simultaneously, however, we can quickly formulate a critique of how far this model falls short of the richness of our language behavior. We know very well that the state of brain and body can modify the process of language production in subtle ways, not only in deciding what indeed is the salience of something in the scene, but also in producing the tone of voice that often says more than words. When we describe something we do not simply give a neutral description of the scene. There is a communicative intent. A beautiful example recently of using words to have one publicly accountable message while saying something different may be found in the book, *The Emperor: Downfall of an Autocrat*, by the Polish writer Ryszard Kapuscinski. It is ostensibly about the late Abyssinian emperor Haile Selassie and his downfall, but I am told that the book may also be read as a parable about the Polish leader Gierek, and his autocratic rule and downfall. Because the book is couched very carefully as a quite factual story of life in a far away place, it was free of the restrictions of the censor and thus was available for other Poles to read knowingly and to understand the implicit critique of their regime.

When we speak it is not only in terms of trying to express the most salient aspects of some semantic network, but in awareness of a social setting, perhaps even seeking to communicate different things to different people. Or an utterance may be a lie. Conversation can be stymied by a misunderstanding of the deep sort that grows out of a radically different stock of schemas. Moreover, language is not only for communication. It is part of the way in which a person alone can make his thoughts available to consciousness, to work on them, play with them, to create new thoughts from them. A good deal of our language activity is internal, and one of the themes of chapter 3 will be to further ponder the role of language as we confront a schema-theoretic view of consciousness with a Freudian view. Such complexities may challenge us to make new contributions to cognitive science or convince us of their futility. In either case, we can agree that the humanities have much to tell us about language that is not now ripe for scientific analysis.

We introduced this chapter with a view of language as expressing the inexpressible. Paradoxically, though, language has the dual role of providing firm anchor points, ways of providing repeatable structures that let us develop and reshape an argument, either alone or in conversation. Perhaps, then, we resolve that question mark that we left when we quoted from Nathalie Sarraute who, having attacked language for its inadequacy, nonetheless said that there is "something in search of its language which cannot exist without its language."

We talked about logic in chapter 1 and saw that to give a full account of human behavior via a logical model would be unduly limiting; and yet there are situations in which there is no doubt that we proceed by a logical analysis of alternatives. In the same way, we must live with the tension between the view that language is, in one sense, impoverished, forcing patterns that can constrict and distort the nuances and subtlety of feeling, and the view that it can, in other cases, allow us to penetrate into realms of thought that would otherwise be denied to us.

Language as Metaphor

Our view of language as providing partial expression of a richer network of schemas, which is in turn the partial crystallization of the true richness of lived experience, leads us to the view that there is a sense in which all language is metaphorical.[5] In some approaches to language, the "literal" and the "metaphorical" form a very strict dichotomy. A literal meaning is somehow the "real" meaning; each word is viewed as having is own meaning, or perhaps a finite set of different literal meanings. To find out what a sentence means, one then has to compose those individual literal meanings. In this approach, metaphor is seen as somewhat aberrant, an exception to be found in poetic language. Somehow, a metaphoric meaning is then a distortion of literal meaning and is only to be sought when the literal rending of the sentence fails. But what we would like to suggest from our schema-theoretic point of view is that meaning is extracted from a dynamic process which is virtually endless. A sequence of words is always to be seen as an impoverished representation of some schema assemblage. When we engage in conversation, we engage in a process of interaction which may tease out more and more of the meaning rooted in this assemblage, and may even change the meaning as the conversation proceeds (whether that conversation is an actual dialogue, or an internal "conversation" in one's own exploration further and further into the network of one's own internal meaning).

Certainly, then, a sentence may have a "literal" meaning in terms of skimming off the most common set of associations from the related schemas. But there is no dividing line that sets off the literal from the metaphorical. In all cases, the schema or discourse provides an entry into a schema network. Perhaps "literalness" is then simply a measure of the speed with which we can break off the exploration. When we try to interpret a poem, we may see ourselves as going indefinitely deeper into this network of meaning, both in articulating the knowledge implicit in the schemas that provide the core of the interpretation and also in exploring the network of associations that takes us further and further out into the richness of experience.

Consider the sentence, "The sky is crying." There is no fixed dictionary that includes literal meanings of "sky" or "crying" that yields a coherent interpretation of this sentence. Rather, each word has a network of meaning, and these networks can interact until a coherent reading of the sentence is found, a reading that can be accepted as is, or elaborated even further. If we start from the "literal" meaning of "sky," then in making sense of "crying" we must strip away those meanings that literally require eyes to be present and perhaps come to focus on the falling of moisture. We might then perhaps come to the notion of rain as teardrops falling from the sky, but, because of our other associations with crying, we can go further into the network to see "the sky is crying" not simply as a funny way of saying, "it is raining" but as enriching this bare statement with a mood of sadness. On the other hand, suppose the context is such that it is the "literal" sense of crying that is anchored. If we think of the sky as normally blue, we might come to see the sentence as asserting that the person who is crying has blue eyes. However, if we let our two metaphors interact, so that the literal sense of "crying" is combined with the metaphorical sense of "raining"—if it really were sky that was doing the crying, the crying would be rain, so that the color of the sky would be gray—the metaphor would now tell us that the person has gray eyes, the gray of a leaden sky, not the blue we normally associate with sky.

With this, a schema-theoretic view of language leads us to see metaphoric meaning as in some sense normal. This is because we view language as embedded within a changing and holistic schema network. We see literal and metaphorical meaning as lying on a continuum, rather than as forming a strict dichotomy.

Our understanding of this continuity can allow us to extend the field of our science as we enrich our vocabulary for talking about that to which science cannot yet do justice. As we extend our perspective, we will come to look at science as something not limited to the purely physical sciences, even in the extended form of the information-processing sciences which include schema theory. Rather, it extends out into the moral sciences and the *Geisteswissenschaften*, with a continuum between those studies anchored in the quantifiable and those anchored in the richness of lived experience. We must continue to explore the creative tension posed by the apparent incompatibility of these extremes. In particular, we may come to view myth not as being dismissable as simply non-science or pre-science but as a rich way of placing lived experience in a meaningful perspective. This is relevant to our attempt in the next chapter to understand what happened as Freud made the transition from a neurologist to the creator of one of the most powerful myths of the twentieth century.

However, I do not want this openness to dialogue to be seen as suggesting an acceptance of the view that physical science is "just one more myth." Mere myths cannot build jet planes that fly, or probe the properties of matter to enable us to build electronic circuitry. Science is driven by the pragmatic criterion of successful prediction and control. To what extent is mathematics a science? Mathematics has criteria that preclude arbitrary statements. In some ways it is like a game: a proof must follow certain rules, and once the rules are in place, one can objectively check whether the rules are being observed. But mathematics is a human enterprise, and it is not enough to obey the rules. We bring aesthetic criteria to bear. To simply generate a statement that follows logically from the axioms does not ensure a publishable paper or the affection of one's colleagues. There are social criteria as to what is a strong result, what is a hard result, what is an interesting result. Another constraint is that even though mathematics is not an empirical science, it provides the tool for many sciences. In applications, the test of the mathematics is not only that it meets the criteria of playing by its rules, but also that those rules are pragmatically successful in that with them we can get a better handle on experiments, and make better predictions.[6] The history of mathematics is rich in the interplay between the joy and aesthetics of the formal game, and social consensus on what are the interesting problems, and the pragmatic sucesses in using mathematics to control a scientific investigation. Mathematics has that element of a free human creation, and yet there are many places where it connects with the pragmatic criteria of scientific research.

More generally, we can see science as bringing into concordance socially agreed-upon methods of measurement with socially shareable types of formal explanation, all making sense to the individual who tries to measure experience against these standards. These general criteria still apply when we move to the less quantifiable concerns of cognitive science. We may expect our person reality to be changed in the light of our attempt at a careful science and our science to be changed and extended as we try to assimilate more of our human experience within the continuing dialogue of two-way reductionism.

Chapter 3 Freud: From the *Project* to Projection

WE HAVE SAID that there is an everyday reality of persons, things, and society and have asked the question, "How can we come to know that reality?" In answer, we have embarked on an attempt to build schema theory as a naturalistic, well-articulated theory of the mind, suggesting that we can understand the complexity of our minds in terms of a network of schemas of which perhaps thousands are active at any time. Those schemas are seen as competing and cooperating to form coalitions, or schema assemblages, which will control behavior, melding our current needs, our motivations and our cognitive representations to actively assimilate what is going on around us at the time. Perception is active, so that it is our current stock of schemas that determines what we will take from the environment. Schemas viewed as units of representation of, or interaction with, the world, can be at many different levels. A schema not only provides abilities for recognition and guides to action, but—and this is the crucial, open-ended aspect—it sets expectations. And, of course, these expectations may be wrong. Our schemas and their connections within the schema network can change; we can learn. Each of us has a different genetic constitution, different life experiences and, thus, different schemas, so that each of us has constructed a different world view that each of us takes for reality.

We have come to see schema theory not as a patent set of answers to our concern with our reality but, rather, as a science in a state of becoming. Schema theory is an open-ended subject that will respond to, but also change, our personal and social realities. And what is this personal reality? Surely at the core of it is the knowledge of our own self-awareness, our consciousness. So I want to now address the problem of consciousness, doing so from three different perspectives. First, looking at the notions of the neurophysiologist John Eccles who sees consciousness as residing in a mind separate from the brain. Second, giving my own view of how consciousness might be seen as an evolved aspect of schema activity; and thirdly, initiating the main theme of this chapter, looking at Freud's view of consciousness in its balance with the unconscious, and the dynamic role played by repression.

The Three World Model

Eccles is both an expert on the analysis of the brain and yet an unabashed dualist; in fact, he is a trialist. He has embraced Karl Popper's suggestion that there are not just World One, the physical world, and World Two, the mental world, but also World Three, the world of the social and the normative (Popper and Eccles, 1977). Eccles argues that the brain is part of World One (it is physical), but that it has a part called the liaison brain that can communicate with World Two, the mind. World Three, for him, is embodied in artifacts, books and other physical aspects of culture, so we communicate with World Three through World One, the physical world. Thus World Three only reaches our mind, World Two, through the liaison brain. Eccles has no notion of World Three as having a direct communication path to the minds of individuals, as one might posit if one accepted Jung's notion of the collective unconscious. Nor, as one might expect given the roots of his dualism or trialism in his Catholic boyhood, does Eccles offer any notion of what we might call World Four, the world of God. This world would provide an even higher level of abstraction from the physical, with a direct pathway from World Two whereby minds could have communion with some greater reality.

Eccles' notion of the liaison brain is influenced by studies of split-brain patients who have had the corpus callosum, which joins the two hemispheres of the cerebral cortex, severed (Sperry, 1968). Studies of such patients, and of patients with one hemisphere removed, suggest that only the left brain has a rich repertoire of speech behavior, although there are some aspects of word use that seem to be available in the right brain. This suggests two possible views: one is that consciousness is only in the left brain; the other is that the left and right brain both have some form of consciousness, but it is the left brain that has the speech mechanisms whereby this consciousness can be communicated. Eccles opts for the view that only the left brain is conscious and then claims that this is so because it can communicate with World Two. He does not inquire whether the structure of that brain could provide a physical underpinning for consciousness.[1] His views can be well seen in the following quotation:

> I believe that my genetic coding is not responsible for my uniqueness as an experiencing being nor do my post-natal experience and

> education provide a satisfactory explanation of the self that I experience. It is a necessary, but not sufficient, condition. We go through life living with this mysterious experience of ourselves as experiencing beings. I believe that we have to accept what I call a personalist philosophy that central to our experienced existence is our personal uniqueness. (Eccles, 1977, p. 227)

Now I certainly agree with Eccles that we experience a sense of personal uniqueness and that our genetic heritage and experiential upbringing provide us with many points of difference from other individuals. Where we would seem to part ways is that Eccles, seeing the tension between the richness of personal experience and what we can currently explain in terms of the brain, would posit a sharp cut between World Two and World One; while we would try to see if we can use this tension to expand our understanding of the mind and its embodiment within the brain.

Schemas and Consciousness

Recall the discussion of two-way reductionism from the first chapter. In the history of science, given two sciences of different aspects of reality, such as the detailed physics of molecular interactions and the thermodynamics of bulk properties of matter, we did not see a strict reduction of the higher-level theory to the laws and techniques of the lower-level theory. Rather, there was a process of dialogue in the course of which both sciences were changed, to yield a new, larger, understanding of the properties of matter. And so, as we confront the richness of mental vocabulary with the growing richness of schema theory, both our scientific theory and our understanding of ourselves will change. With this perspective, let us see how a conscious mind might emerge as embodied within all this schema activity.

We each have only one body to act with and thus have a limited set of actions available to us at any one time, so that, as we move towards the actual commitment of the organism to action, there would have to be a channeling from the richness of understanding to the well-focused choice, not necessarily conscious, of a course of action. Large ensembles of schemas in some sense interact, compete, cooperate to constitute a relatively well-focused plan of action that will commit the organism. This combination of many schemas with one body suggests both a continuity of behavior by the one individual in similar situations, but also, as this repertoire builds up over time, the possibility that the schemas may eventually cohere in new ways, so that what had been an expected behavior in a certain set of situations may eventually give way, through new patterns of schema interaction, to new courses of behavior.

Where we stress schemas as units of process and representation, Piaget looks at a schema as a pattern of regularity in the overt behavior of the child. As he follows the child through its cognitive development, he sees the child as going through successive *stages*, or styles of schema (Brainerd, 1978). He sees the progression through the stages as driving the process of schema change. I would rather see the stage as a pattern of coherence that can be mapped across the many schemas that constitute the action and perception of the child. Then, through the cumulative local changes of individual schemas, there can be a sort of phase transition which can be *described as*, not explained by, a stage change. In science, Kuhn's paradigms (Kuhn, 1962) can be treated similarly. In my individualistic view of science, there is no transhuman process that suddenly decrees that a science will flip from the Newtonian to the relativistic mode as an external controlling reality. Rather, scientists bring to their study a body of techniques, a body of facts, a body of theories. They begin to build up a certain number of new models as discordant facts are expressed within the current observation level. And then somebody discerns a new pattern, and this—to use a metaphor from crystallization—provides the seed whereby a sudden change can then propagate through the individually held schemas. What we see then is that each individual has sets of schemas with some sort of coherence between them (this is not to claim that *all* of an individual's schemas are coherent); and the style of such a set of schemas can change over time to provide some sort of unity.

I speculated that there are perhaps hundreds of thousands of schemas corresponding to the totality of what a human knows, with perhaps hundreds of these active at any time to subserve the current interaction of the organism with its environment. By contrast, consciousness seems to be rather focused, with perhaps seven or so schemas in direct awareness at any time.[2] But what is the role of consciousness? Wouldn't these schemas do their jobs just as well if there were no such thing as consciousness? I cannot make a case for why we must *be* conscious. I think I can make a case for why we have patterns of schema activity that *correlate* with consciousness very well, and I will leave it to the phenomenologists to take the matter further. But first, I do want to stress that our common sense about consciousness can be misleading, that there are many things that we think require our conscious thought that do not. A very dramatic example is the phenomenon of blind-sight.[3] When a patient who has lost his visual cortex, but still has the mid-brain visual system intact, is asked to say where an object held up in front of him might be, he will say he cannot see it, he has no conscious knowledge of where the object is. But then you say, "I'm going to wave it in front of you and although you can't see it, please humor me and guess whether it's on your left or your right," and the person guesses correctly. Or you throw him a ball and he catches it. Actions we would normally think of as requiring our conscious awareness do not always do so. The schemas manage without the direction of consciousness. Why, then again, consciousness?

I don't know, but my story, my myth if you will, of consciousness starts from the nineteenth-century observations of Hughlings Jackson.

Hughlings Jackson was an eminent nineteenth-century British neurologist who was very much influenced, as many people were at that time, by Darwinian concepts of evolution.[4] He tried to explain the effects of many brain lesions in terms of removal of the more highly evolved aspects of brain function, so that what he saw in the performance of a person with a lesioned brain was the expression of evolutionarily older mechanisms of the brain. We may think of the brain as evolving in such a way that new brain regions or new schemas can exploit the prior richness of the brain. But the crucial point, and this is very much part of Jackson's analysis, is that once these new regions are in place or these new schemas are available, they provide a richer environment for the older parts of the brain. These now have new possibilities for further evolution, whether evolution of brain regions over a biological time scale or the evolution of schemas over an individual time scale.[5] If you compare, for example, the brain of a frog with the brain of a cat, which we might think of crudely as being snapshots of rather different stages in evolution, we find that the corresponding regions of the visual mid-brain are much richer in the cat, not so much in terms of retinal input, but because there are so many descending pathways whereby the new richness of cerebral cortex can influence what goes on in these more classical brain regions.

Perhaps the development of animal and human communication may also be seen in terms of such evolutionary interaction to give us some insight into how consciousness might have evolved. One of the very striking features about human ability is that we come to incorporate tools into our body schema. When we use a screwdriver our body ends at the end of the screwdriver, not at the end of the hand; when we drive a car, our body ends at the rear bumper, not at our buttocks. Analogously, as creatures developed as social animals (and this account is not restricted to humans), the body might end not at the extremities of the physical body, but extend to incorporate aspects of other members of the group. However, coordinating others is a more subtle matter than just directing an arm or slightly adapting the hand to control a tool. The social animal has to find a way of expressing some précis of its mental state, of its richness of schema activity, so that it may then impinge upon others so that their behavior may be coordinated. With increasing richness of social interactions, though still at a prelinguistic stage in our evolutionary story, there would come the ability to form a précis of schema activity that is not necessarily relevant to deciding what to do next, but is relevant in terms of coordination with others.

Then, adapting my Jacksonian theme, I would suggest that at the first stages of its evolution, the "précis" just serves for communication. It is a gloss on the richness of activity in the schema network. But once it is available as part of the way the brain has evolved, the older patterns of schema interaction, which were relatively unconscious, become modified. That précis is in their environment. There are occasions on which they can be better coordinated if they coordinate via the précis than if they coordinate by themselves. We then have a subnetwork of the schema network which provides a précis that may often have no directive role, and yet may evolve to have a role that is, sometimes but not always, directive. This evolutionary process, which occurs with subhuman species, then sets the stage, I would suggest, both for human consciousness and also for the evolution of language to express this rather coarse précis of the richness of the underlying schema activity. In summary, this pretheory sees consciousness as a précis of schema activity, evolving in such a way that it can elaborate certain mental processes at the level of language and logic and is related in part, but not entirely, to communication.

Freud and the Unconscious

The Freudian story[6] about consciousness is not totally discordant with my schema theory, and yet it forces us to address a number of things missing from the hypothetical evolutionary story I've given so far. We started from schema activity and found the existence of conscious processes puzzling from such a perspective. Freud starts from a view of the mind as conscious. He then adds the notion of the preconscious as a store of ideas which are not currently active, but latent, ready to be used when the need arises. Freud's essential contribution was not simply to observe that there are things that are unconscious (this had been observed many times before), but to introduce his notion of repression as a dynamic process. He postulated that there are ideas so heavily charged with sexual and aggressive energies that to bring them into consciousness with the consequent flood of associations is so painful that the ideas must be repressed. Freud has an associationist model of the mind, which he sees as made up of ideas, each associated with other ideas. In particular, even repressed ideas will have associations with other ideas. Freud thus sees people as repeatedly in the unfortunate situation that, to avoid a repressed idea itself, they must avoid many other ideas with which it is associated. Notice how rich this makes the unconscious. It has the mechanism to recognize that, in the network of ideas, it is approaching ideas associated with a danger point and to steer the associations away. But of course this must distort thinking. In extreme cases, the patient may find himself unable to act simply because the wherewithal, the stock of ideas necessary to come up with effective action, is blocked.

Freud sees the conscious as coextensive with the linguistically expressible. The repressed is unconscious, but may well have been linguistically expressible prior to its repression. The key idea of psychoanalysis, expressed rather crudely, is to offer a word therapy whereby one can use linguistic anchors in the conscious network to begin to probe further through the network of associations to bring certain ideas from repression into consciousness. The hope is that, having brought them briefly into consciousness, one can then analyze them, discharging their associated negative energies to dissect them out of the network of repressed ideas. As Freud said, the process does not bring happiness to the individual, but at least brings the patient to a state of normal unhappiness rather than the pathological unhappiness that marked him beforehand.

Our schema theory offers a way of going beyond a crude associationism to a more subtle notion of the interplay of ideas, which are now seen as rich schemas. It also suggests the futility of the view that all of the unconscious can be brought into the conscious, even if Freudian therapy can succeed by bringing certain trouble spots into consciousness where they can be diagnosed and the distortions they induce diminished. Through such considerations, the study of Freud can help us to move beyond the emotion-free schema theory of the previous chapters, even if we have reservations about the emphasis on sexual energies or about certain mythological expressions of the Freudian theory. There is a certain irony here. In the last two chapters we have tried to enrich a logical analysis of the mind to come up with a schema-theoretic analysis, and this approach faces difficulty in analyzing the emotions. But if one comes to the problem from the side of person reality, it is the emotions that are primary, and what needs explanation is how logic can play a role in the human condition.

From Neurology to Mythology

Let me turn from consciousness to a consideration of the transition in Freud's intellectual career from neurology to mythology. How is it that the early writings of Freud are securely anchored in the neurology of his day, and yet much of what we remember about Freud is couched in the language of the Greek myths? Does Freud's turning from brain to myth suggest that we should all abandon too scientific a world view and instead seek to understand personal reality in mythological terms? Can we, as a two-way reductionist might argue, see person language as a convenient level of description for what is nonetheless the working out of the neural complexities or must we see the person as transcending what any physical naturalistic theory could hope to account for? But first, let me trace the history of Freud's intellectual development, even if with disconcerting brevity.

In 1891, Freud published a monograph, *On Aphasia*, which unfortunately is never reproduced in any of the standard editions of his work. As a neurologist, Freud surveyed work going back to the contributions of Broca and Wernicke on localization, studies of the extent to which one part of the brain could be seen as being responsible for speech production, another as responsible for speech perception, and so on. He gave a critique of overly localizationist ideas, not unrelated to our current views of cooperative computation in which most processes are to be seen as involving the cooperation of different brain regions rather than the localized activity of one brain region.[7]

In those quaint, ghoulish days before CAT-scans (see Kertesz, 1982 for an exposition), the neurologist waited until the patient died and then dissected the brain to try to find lesions which could be correlated with the form of pathology. Freud was greatly influenced by the study of hysterics, who exhibited symptoms that would seem to make them patients of the neurologist rather than any other specialist, but who exhibited no identifiable lesions at postmortem. Again, when a stroke affects the brain, the nerve tracts are seen to map the body onto the brain in certain ways. But when a hysteric came to the neurologist complaining of numbness in part of the body, the area of numbness did not correspond to an area that could innervate part of the brain. Rather, the hysteric had a numbness that ended at the elbow, say; a numbness that involved her body schema rather than the pattern of neural innervation. These observations led Freud, who had been brought up as a nineteenth-century materialist in the tradition of Helmholtz, to the reluctant conclusion that the lesion was in some sense psychic rather than neurological. However, to maintain some materialistic biological respectability, he did give an essential place to the instincts, which he saw as providing the bridge between the psychic representation and the physico-chemical forces of the body.

With the publication of the *Studies of Hysteria*, 1895 becomes a very important year because it marks the transition from Freud as neurologist to Freud as psychologist. But 1895 is also interesting for those of us who want to understand Freud's rootedness in neurology because it saw the writing of what is now called the *Project for a Scientific Psychology*. This draft of a neural net theory of the mind was sent in manuscript to Wilhelm Fliess but was not published until 1950.[8] Thus, prior to 1950, most students of Freud did not understand that, as is clear from the *Project*, many of his ideas such as libido come from a particular theory of how neural networks function in the brain. The years around 1895 are important in the history of brain research, for this is the period when it was still controversial as to whether the brain was one totally interconnected mesh, a syncytium of cells all connected together, or whether it was made up of discrete elements, neurons. The neuron doctrine won, thanks to the work of Ramon y Cajal (1937) at the turn of the century. Freud was convinced of the cell doctrine of discrete intercommunicating cells, but his view of the brain preceded the electrochemical models (Shepherd, 1979) that now drive our thinking. He had a hydraulic view of the brain, akin to Lorenz's hydraulic model of drives in ethology (Lorenz, 1981). Freud saw each neuron as having a charge of energy, and the job of the neural network, which was continually being charged by the stimulation of the senses, was to find ways of discharging that energy. He then looked at mental pathology in terms of what could block the discharge of the brain. In the *Project*, then, we see the intermixture of neural network ideas with his notion of the ego and the unconscious and the discharge of libido.

When we move on to what many consider to be the canonical introduction to the mature Freud, namely *The Interpretation of Dreams* of 1900, we find that neural nets are not mentioned. Yet the basic use of such terms as cathexis and libido are best understood in terms of the energy flowing amongst the neurons in the networks of the *Project*. However, *The Interpretation of Dreams* is not only interesting because of the way in which neurological notions have been sublimated into the familiar Freudian terminology, a purely psychological terminology, but also because in chapter 5 of that book, we see the Greek myth of Oedipus introduced to help talk about the human condition. In discussing this myth, Freud comments how powerful it is as compared to the contemporary tragedies of the Viennese stage, and he tries to understand this power of the Oedipus myth and of Greek myths in general. He concludes that it is because the myths embody basic motifs which resonate with human experience.

An interesting transition occurs in Freud's writings. In *The Interpretation of Dreams*, the Oedipus myth is just that, a myth which is part of the culture that a civilized member of the Western world would have gained from his education in that day. The

Greek myths provide a powerful account of the human condition, but they have no reality status beyond that. However, as Freud's work progresses, he moves from Oedipus as a myth resonant with aspects of infant sexuality to the Oedipus complex as an explanatory principle for the development of infantile sexuality. The constitution of our human brains is seen to be such that we will inevitably go through this psychologically real structure of the Oedipus complex. *The Interpretation of Dreams* sees the attempt to turn neurological language into a useful psychological language; later writings turn from the realization of the power of certain Greek myths to their embodiment in Freud's metapsychology, and the creation of what we might call the Freudian myth. In hermeneutic terminology, we may talk of a merging of the two horizons of neurology and mythology. As in our two-way reduction, neither viewpoint emerges unchanged. Neurology is purged of its detailed analysis of individual neurons, but has provided a psychological language. The myths have moved from treasures of a distant Greek culture to actual causal principles in human development. We see this not only in the Oedipus complex, but also in the way in which the biological drives become reexpressed in terms of Eros and Thanatos, mythologizing the biological drives. Freud came to see a process of growth from the pleasure principle of the young child to the reality principle, and this included the ability to face the reality of death. He believed that it was only in facing up to that reality that one could move beyond the pleasure principle to become a fully alive human being. The humanist might see Freud's turn to mythology as further evidence for the impossibility of a scientific and naturalist account of the mind. A scientist might respond that Freud did not have access to modern neuroscience, and so gave up the search for a neurological theory of mind prematurely.[9] My stance, as a two-way reductionist, is to reiterate that we do not have a complete science of the mind, yet to also stress that the quest for such a science has no established cutoff. And so we work in the tension between our schema theory and the intensely creative descriptions of the person that we find in literature and Freud, and that we cull from our own lives.

To bolster the claim that there is no limit to the mental above which physical explanation must break off, I will discuss some of the notions of the Oedipus complex from a more schema-theoretic point of view. In his book *The Ego and the Id*, Freud returns to the Oedipus complex, but now concedes that the young male, rather than being dominated by the Oedipus complex of love of the mother and a death wish for the father, instead comes to exhibit a twofold Oedipus complex, as do all children, male or female. This involves both the boy's love of the mother and the girl's love of the father and the consequent tensions. From a schema-theoretic point of view, I would respond to this twofold pattern in a somewhat different way, asserting that the young child develops schemas for the mother and the father. As part of our challenge to schema theory, we must now distinguish schemas for observing something from schemas for being or doing something, as in the difference between a schema for recognizing an object and the schema for exhibiting a skill. As the child observes his or her parents, he or she comes not only to learn to recognize them, and certain of their behaviors, but also to interiorize them, to form schemas for acting as the parent does. This is our schema-theoretic form of what Freud would call *identification*. And then we come to a theme that recurs throughout Freud, that of the tension that leads to repression. Having begun to build schemas for acting like the parents, the child discovers (not necessarily consciously) that certain parental activities are not allowed to the child. The building of identification is in tension with punishments or sanctions that block certain parts of that identification, a tension that may or may not block the development of a coherent self schema. This schema-theoretic view of the mental development of the child addresses a number of Freud's observations, but lets us begin to understand why the reality of the acquisition of schemas might indeed resonate with the Oedipus myth without leading us to postulate the Oedipus complex as a psychological reality.

Note the difference in "grain." We stress the dynamic interaction of many schemas where Freud emphasizes the admixture of the male and female forms of the Oedipus complex as affected by a few basic instinctual processes. Piaget emphasizes the stages the child must go through, with the child's individual experience or social environment changing his development only by not providing sufficient aliment to allow the child to proceed through the full sequence of stages seen in the Swiss child. My more detailed view of schemas suggests that the views of Freud and Piaget are still too phenomenological, too classificatory. Schema theory, as I visualize it, tries to move down a level, to the richer notion of a stock of schemas of which a substock is deployed in our activity at any one time. Then we try to understand patterns of coherence and patterns of tension within these schema interactions to address what it is that we take from Piaget, what it is that we take from Freud. But to the extent that Piaget sees predetermined stages, and to the extent that Freud sees us determined by a rather small stock of complexes and instincts, we must move beyond their work to address in greater detail both the individuality of each person's life and the diversity of social structures and, therefore, the very richness of the interaction between individuals and society.

In artificial intelligence, we sometimes distinguish between knowledge encoded as programs or chunks of data, and the control structures whose job it is to sit above those units and manipulate them. In some AI systems, however, the control structures are themselves embodied in active pieces of knowledge and the entire system operates through their inherent interactions. In a computational theory of the mind, it can thus be a real issue as to whether there are control structures separable from the schemas, or whether, in fact, all the control structures are themselves distributed across schemas. This question pervades my current thinking about concepts like drives (Lieblich and Arbib, 1982). At times I might want to talk as if there were a schema for each particular drive; at other times I would want to see a drive as part of the underlying economy of the control structures. However, the fact that we have a word to describe human mental states does not imply that there is a schema or control structure described by that word (as distinct from a schema for that *word*, a schema which allows us to recognize a situation in which we wish to utter the word or recognize the spoken word and incorporate it into our ongoing model of a conversation). Nathalie Sarraute wrote that she would not trust the words "happiness" or "enjoyment" or even the simple word "joy" to describe the richness of a childhood experience. Even though we may use the word "happiness" to provide a crude language-level caricature of a range of experiences, there may be no actual motive or drive of happiness as an operative principle. Again we have the notion that a schema can be something we posit as an internal structure which is playing a causal role in our expression of behavior, or may be just a way in which we observe a richness of behaviors to arbitrarily carve out and label a subset, seeing regularities that do not express any internal or social reality.

By confronting Freud's account of the complete Odeipus complex with our own schema theory, we see that an adequate theory of schema formation and interaction must include schemas for person-centered affective skills as well as sensory, motor, and linguistic skills. Our notion of schema networks then reminds us that schemas cannot be neatly classified into these distinct categories.

Superego and Society

Freud has a spatial view of the mind as comprising the distinct structures of the id, the ego, and the superego (or ego ideal). He answers the question of how the individual becomes a member of society by saying that the superego provides the mental representation of society. The superego is, in some sense, the dressing-up in psychoanalytic terminology of the old-fashioned notion of the conscience, but Freud gives an account in terms of his metapsychology by basing the superego on the first identification with the parents. In contrast to the orderly Piagetian progression of reflective abstractions to better and better generalizations, the Freudian account sees no progress without an accompanying and increasing burden of tensions. The blocking of certain aspects of identification with the parents becomes the basis of what society says one can and cannot do, yielding the tension, experienced in the sense of guilt, between the demands of conscience and the actual attainments of the ego. The individual's schemas of society, of which we spoke in chapter 2, do not quite correspond to the superego because for Freud the superego is charged with guilt. The superego is in some sense a "bad thing" in that it provides a repressive structure to be worked through into the ego so that it can be overcome to yield a less unhappy mental life. This is not unlike the Marxist view of ideology as a purely repressive structure imposed upon individual people, limiting their freedom rather than providing necessary supports for their social development.

We contrasted the individual's schemas of society with social schemas in the sense of the regularities expressed in social behavior which constitute an external reality for individuals. We did not see those social schemas, those external schemas, as codified in any one place or expressed in any one mind. Rather, it is the coherence of a society with its many different actors as perceived by the individual that can provide the external structure that can be codified in the individual schemas. Our general ability to discover patterns in overt behavior leads the individual to form schemas representing society or embodying social behavior, but because these individual patterns are based on different partial samples of societal regularities, they can be in discord with what is demanded by other members of society. And so we come to another view of the tension between self and society, which looks in somewhat more detail at the way in which the schemas in the head are built up, not only in terms of the immediate tensions of the primal identification, but also in terms of the overall pattern of interaction throughout life. In some cases the order perceived in the complexities around one gives one the aiblity to interact happily with the world; in other cases, it leads to tension.

Religion and Civilization

We have seen how, for Freud, the Oedipus myth became the psychological reality of the Oedipus complex. By the time he wrote *Totem and Taboo* (1912–1913), Freud had come to view the Oedipus complex as a historical reality, rooted in the history of mankind, when the primal horde murdered the father. It was the consequent mixture of guilt and veneration, Freud says, that combined to yield the basic elements of religion. This story is not supported by current anthropology; Freud has created his own rich myth of human history.

People have a diversity of religions and of concepts of God. Freud's myth addresses the myth of God the Father, as distinct from, for example, the myth of God the Creator. He came to see God not as a transcendent reality beyond space and time, but rather as the projection of the Oedipus complex. He believed that people, in trying to make sense of feelings of helplessness, of dependence, of guilt or veneration, projected this outside themselves, creating this new reality of God to act as the target of their many feelings as they tried to wrestle with the complexity of life and society. Freud sees this God as an illusion. In his 1923 critique of religion, *The Future of an Illusion*, he distinguishes illusion from error or delusion. An illusion may be true, but a belief becomes an illusion when wish fulfillment is the main factor in its motivation, rather than a regard for its relation to reality. Through his myth of the primal horde, Freud suggests that, to use our terminology, we acquire our God schema not because there is objective evidence but because we wish that there were a father who could handle the cosmic helplessness we feel in the way that our own father was able so often to minister to the helplessness that we felt as a child.

Freud's resistance to illusion was combined with a deep pessimism. He accepted that many people would need the comforts of religion, but he could not accept them for himself. In particular, it seemed to him that aggression and guilt could not be eliminated from human life. This feeling was reinforced by the disillusionment felt in the West in the wake of World War I, for the war showed that the barbarian was still with us. The myth of progress of the nineteenth-century was now revealed as an illusion, a wish fulfillment. In terms of Freud's own history, we know that later he was under house arrest by the Nazis for two years and that two of his sisters died in concentration camps. His pessimism (or shall we call it his realism?) that sees aggression and guilt as integral parts of the human condition are reinforced for us by what we now know has happened since Freud's death.

In *Civilization and Its Discontents* (1930), Freud saw, as corollary to the inevitability of human aggression, that civilization must demand renunciation of certain of our instinctual gratifications. He saw civilization as always being precariously balanced between social norm and instinctual demands, and thus as being ambivalently regarded because of the tension between the pleasure of "cutting loose" and the sober appreciation of the dangers of such abandon. We are back to the importance of repression. To function happily within society, one has to repress certain parts of one's being. Freud again reminds us of the power of instinctual schemas. Whether or not we are prepared to see in them the Manichaean struggle of Eros and Thanatos, he has given us the concept of repression and linked it not only to individual psychopathology, but also to the very foundations of civilization.

Where Piaget gave us what now seems an almost utopian vision of reflective abstraction yielding coherent networks of ever greater generality which can integrate and illuminate our more concrete schemas, Freud reminds us that consistency is not an attribute of humanity. In fact, our schema theory places as much stress on competition between alternative schemas as it does on cooperation between those that can cohere within a particular situation. Freud tells us this is not simply a computational abstraction, but is part of what we must confront as we come to try to understand the darkness of the human soul, a very real part of the human reality.

Freud certainly accepts that most people may need illusion, seeing religion as expressing in symbolic form the human feeling of helplessness. He concedes that it offers rituals that may help control and give meaning to man's innate aggression and sense of guilt. It is important to note, then, that Freud's atheism does not detract from his understanding of the social functions of religion even though Freud would himself forego the comforts of illusion and look to science for what limited progress may be possible. The German theologian Hans Küng (1979), in a little book called *Freud and the Problem of God*, seems not to understand Freud's point. He says: "After the experience of National Socialism and communism, modern atheism has lost much of its credibility. The ambivalent character of progress in science and technology

. . . [leads] many to doubt that science . . . is . . . the source of the universal happiness of mankind which, according to Freud, is not provided by religion." But, of course, Freud did not hold that science promised universal happiness either. He just denied that alternatives to science would be more successful.

Chapter 4 Schemas and Social Systems

IN THIS CHAPTER we probe further the social dimension of knowledge, asking in particular what kind of knowlege we can have of people, their values, and their ideologies. We shall address this question at two different levels:

The first is that of the community of scholars who want to understand the nature of social science. What is it that we as investigators, as reflective critics, can learn about society? We agree that the social sciences are different from the natural sciences. Nonetheless, I suggest that there is a unifying methodology which will allow us to relate the natural sciences and the social sciences. This unified methodology rests both on Mary Hesse's philosophy of science and on the European tradition of hermeneutics.

Over and against the work of scholars in trying to build a reflective scientific analysis of social knowledge, there is the second level, that of individual experience: how is it that we as individuals come to be members of society, and to what extent can we, having become members of a society, provide a critique of it? Here we will build on the discussions in chapter 2 that related the analysis of individual schemas to the external "reality" of social schemas.

The Goals of Social Science

We have seen that the natural sciences are guided (at least in part) by the pragmatic criterion of successful prediction and control: we formulate hypotheses, and we test them by feedback from the external world. We have also seen that the Piagetian theory of assimilation and accommodation offers a similar account of a child's developing knowledge of the world, a world that included not only interaction with the physical world, but also interaction with parents and other actors in the social environment. Can we, in this same light, see the social sciences as proceeding by the pragmatic criterion? This requires some fairly subtle discussion. Many people argue that prediction and control are not the proper goals for knowledge of social and human behavior, but perhaps they view prediction and control in too pejorative a sense. If by control we mean acting in such a way that we force another person to behave in some predetermined way, whatever his prior intentions, most of us would indeed regard this as abhorrent and feel it would not be the proper aim of our understanding of society to achieve control in this sense. But if by prediction and control we mean being able to choose a course of action that will make a friend happy rather than sad, as it might properly be in the human domain, then the pragmatic criterion does indeed seem relevant both to social scientists and to our own moral development.

Rather than seeing sciences as fixed objective bodies of knowledge, Hesse (1980) sees natural science as engaged in a continual dialogue between observation and theory. There are two languages in interaction, the *observation language* in which we express our observations about the external world, and the *theory language*, in which we express our formal accounts of those observations. The crucial idea is to see the observation language as being itself theory dependent. For example, a cyclotron yields very abstract measurements which we take to tell us about the external world. Yet those observations cannot be made without accepting a body of theory. Moreover, the semantics of the theory language involve sensory-motor schemas. It requires people to agree on what actions and perceptions are involved in deciding that a certain state of the external world obtains, that a particular measurement has been made. The theory language is in some sense richer than the observation language because it must involve talk of "hidden" variables that we cannot perceive directly, but which are required to tie the observations together in a richer network of theory. Yet the observation language is itself rich enough to express statements which may contradict the claims of theory. One can state observations that the theory is unable to explain, thus challenging the theory to evolve in turn. Contrast this with George Orwell's NEWSPEAK, the language of *1984*, in which because of slogans like "Love is hate" or "War is peace" it was impossible to state anything which was not tautologically true.

Having seen science not as static but as engaged in a continual dialogue between observation and theory, we can reexamine the social sciences. We have to admit there is no objectifying pragmatic criterion that can bring the social sciences to the same objectivity as the natural sciences. But what do we mean by objectivity here? It is a contingent fact that there are phenomena well captured by little boxes with dials or digital readouts so that people can agree that certain numbers are correlated with certain conditions, yielding "hard data" as observations; and we might say that the natural sciences are those for which the phenomena are objective in this way, that they can be captured in terms of numerical measurements for which there is an extremely high probability of intersubjective agreement. If this is our definition of "objective," certainly social science can seek objective laws for some of its phenomena. But what can be measured in an intersubjectively agreeable way does not exhaust what the observation language of the human sciences would seek to cover. The observation language for social science must make contact with the unquantified person reality of everyday experience. Defining a belief in terms of the schema that embodies it, we are required by the theory, at least in principle, to give an account of where the schema comes from, either through an evolutionary process that made it an innate schema, or as something the child acquires through experience. If we go back to the child's earliest development, we find that the beliefs about the facts of the world are already tied up with the child's interests to gain food and to socially interact. Schemas thus start by being entwined with value, entwined with interests. There are certain processes whereby we can make these schemas less value laden, so that we can see natural science not as being on the other side of a divide of fact from value, but rather as being a place along the continuum where we come to have relatively objective agreement. Reading a meter is not value laden, but there are observations that are inextricably so laden: "that is pleasant," "that is good," "this is evil." There is no way to come up with a strict divide between facts and values.

A schema was both a process and a representation of some aspect of what we took to be reality; schemas express our apprehension of reality. We emphasized in chapter 1 that a schema is like a computer program in that it is not something of a particular size, for any schema could become a unit in constructing a larger schema. Thus a body of science is itself a great schema incorporating a network of smaller schemas, representing particular theorems, particular experimental techniques, particular texts, and particular critiques. This great schema can serve as an external schema, shared by a whole community, encoded in the minds of many people, in books and social institutions, going beyond the schemas of any one person. An individual internalizes certain aspects of this social schema and perhaps in the process distorts some of the subschemas that constitute that overarching schema.

Hermeneutics

Where are we to turn for the language in which we can think about the dialogue between observation and theory as we move beyond the numerically precise observations of the natural sciences? I shall suggest that a critical, schema-theoretic reflection upon the continental tradition of hermeneutics may well be appropriate.[1] The tradition starts in the 1820s or so with Schleiermacher's attempt to formulate a defense of religion for its cultured despisers. He offers a way of reading the Bible that clears away accretions of the centuries, making it meaningful and understandable to the people of that time. This tradition continues in the work of Dilthey. Kant had asked how one could come to certainty about physical law, and he gave his theory in terms of a priori categories. From our Hessean perspective on the physical sciences, we no longer believe that we can have absolute certainty about physical law, and so no longer accept the notion of a priori categories that will survive critical examination in the development of science. But Dilthey was working in that Kantian tradition and asked what the criteria for knowledge might be in the *Geisteswissenschaften*, the sciences of the human spirit. Schleiermacher, Dilthey, and, later, the English tradition with

Collingwood had the notion of trying to understand a text or historical event by putting oneself in place of the actor of the historical scene or the writer of the historical text to understand the rationale for his action or his writing. The hope was that this approach—empathetic, though apparently subjective—would begin to restore the objectivity of the text as different investigators came to converge on a reconstruction of the original intent, and thus meaning, within the text.

To the contrary, our study of schema theory has taught us that empathy is an idiosyncratic matter which tends to differ on the basis of cultural distance between the observer and the period or culture observed. Max Weber, for example, certainly accepted that when we investigate human society, we do so from within our own cultural perspective and will emphasize what is of value to us in our own society. Nonetheless, he wanted us to try to avoid imposing our judgments on other societies. In that sense, one's interpretation was to be value free, since one was to try not to impose value judgments in one's assessment of what had happened in other places at other times. Nonetheless, social scientists might claim to understand what goes on in a culture more explicitly than members of the culture themselves. For example, we may come to see the functional role of religious beliefs, seeing the way in which they yield family or social solidarity, where for a person within the tradition, the religious beliefs simply describe reality.

All discussions of this kind lead to the problem of relativism. In Kant's discussion of the a priori, and in Dilthey's view of the hermeneutic method, there was the quest for objective knowledge or intercultural grounds of value. The hope for objectivity is a powerful feeling; for example, the hope that given a group of different cultures we could somehow meld them into a harmonious whole. To take a very important question for many people: Could we somehow fuse all religions, including the true teachings of Christ and the Buddha, to come up with the right way of looking at the world? Would this lead to some objective religion or sociology that was less culture centered? I think not. Consider a very specific example: The Judeo-Christian tradition of the soul is based on the notion of the continuing individuality of the self which transcends even death; while Buddhism has a belief in reincarnation which is not seen as a continuation of the individual, but rather as a process of emancipation from the individuality of the material person. It simply is not meaningful to try to merge such different views.

Perhaps we may take a different tack. Just as two scientists with an idea of objective physical reality might agree that their interaction would lead to mutual change on certain matters of principle and a more correct theory, so perhaps we could see an evolving understanding of social systems which yields a process of mutual accommodation in which both social systems are changed into a new more comprehensive system. The person who perhaps expresses this approach to hermeneutics as well as anyone is Hans Gadamer (1976) who sees society and culture as largely mediated to the individual by language and thus rooted in tradition in such a way that tradition cannot be eliminated from the individual's perspective on the world. But, and this echoes what we saw before in Mary Hesse's philosophy of science, that language which mediates the rooting of our beliefs in tradition is itself rich enough to express, or to allow us to comprehend, some critique of the tradition. In particular, it allows some sort of dialogue with people who are in other traditions. Gadamer suggests that we may move toward objectivity if we can engage in a symmetric dialogue. If that dialogue is with someone else, the concept is reasonably clear. If the dialogue is with a text—and much of hermeneutics relates to the meaning of texts—one would simulate this dialogue by trying to reconstruct the surrounding context, the history, the archaeology, and letting it add its voice to that of the text. The aim is to come up with mutual understanding as distinct from a separation of observer and observed.

This differs from the hermeneutics of Schleiermacher, Dilthey, and Collingwood in that it does not aim at recapturing a unique original meaning in some objective way. Rather, there is to be a *fusing of horizons* in which the presuppositions of both partners in the dialogue are shifted, with neither remaining fixed. In the process, both will be changed. To take an example, imagine that the Western view of what constitutes medicine engages in a dialogue with tribal medicine, trying to understand it sympathetically rather than simply rejecting it. The result might be a more positive attitude to the psychosomatic dimension of medicine rather than a drug-centered therapy, yet without rejecting the many contributions of such therapy.

The view of hermeneutics with which we shall continue is the Gadamerian, or *perspectivist*, form—the critique of knowledge which proceeds through a dialogue between different perspectives. Many hermeneuticists start by assuming a split between the natural and the social sciences, or a dualism of things and of persons. By contrast, the strategy adopted throughout this volume is a hermeneutic dialogue that takes things and persons as different horizons which we seek to fuse, as, for example, when we discussed Freud's work as an attempt to fuse the horizons of neurology and mythology in coming to understand the human psyche. We provided a further critique from our own evolving view of the brain and cognitive science to migrate some of Freud's insights from the mythological into what we might find more acceptably scientific. But notice that the science we have now is a dramatically different science from that of eighty years ago. The computer has allowed us to speak scientifically about information; while internal states, which seemed to involve a false subjectivism to the hard-headed behaviorist psychologists of the early years of this century, now have scientific respectability because in the computer we can make an internal state, or a program, an objective reality that we can analyze and understand.

The Ideal Speech Situation

Jurgen Habermas (1971) offers another view of Freud (see also Keat, 1981) by looking at the conduct of psychoanalysis as itself an exercise in hermeneutics. Psychoanalysis is an attempt to unmask psychic delusions to restore the patient to what he regards as a tolerable life. This process may succeed by bringing the patient to accept a story of his childhood which is not necessarily historically true, but which integrates his experience in such a way that he can come to terms with it and regain the state of normal human unhappiness. It is the interaction between patient and therapist that produces this new state. While some psychoanalysts may see their work as bringing the patient back to a canonical state of normalcy, most analysts do not assume that there is a normal state to which the patient is to be returned. They rather seek to create an unpredictable new situation out of what has come to be seen as distortion and delusion. Psychoanalysis is then a nonobjectified science which remains open ended. There is no fixed goal, but rather a continuing process of discovery. We might note here, for later reference, some similarity between Freud's acceptance of this open-ended view of the analytic process and Marx's rejection—though a qualified rejection as we shall see—of essences in utopias and his insistence that man makes himself in the course of the historical dialectic. But with this, I can now turn to the third round, as it were, of the hermeneutic debate. Habermas has sought to generalize Freud's insights to take account of disorders of society as well as those of the individual. In doing this, Habermas accepts the importance of the hermeneutic model of dialogue. But he is critical of Gadamer's approach because he feels that it neglects the social critique of power relations, whether those be unconscious in the Freudian sense or based on the false consciousness of class domination in Marx's sense.

Habermas wants to lay the foundation for objective value criteria from which we can judge historical episodes and he wants to do this by addressing the false consciousness of both the horizons that are being merged. To achieve this, he tries to distinguish communicative action from communicative discourse (Habermas, 1979). *Communicative action* is unreflective and conceals the norms and values of the participants. It is business as usual. Habermas attempts to build what he calls a *universal pragmatics*, which goes beyond the normal syntax and semantics of language to look at how language is used and reflect upon whether it is being used effectively. His idea is that this universal pragmatics is to lead from communicative action, which is unreflective, to *communicative discourse* where critique can operate and ideological distortions can be overcome. He is concerned that power relations may block such discourse, and so he introduces the concept of an *ideal speech situation*. His concern is that Gadamer, by not being concerned with power, did not understand what would be involved in making a speech situation, a hermeneutic dialogue, truly symmetric. Habermas thus stresses that in the ideal speech situation the participants would have equal status with respect to power, such as to issue permissions or commands. There would also be a symmetrical chance to offer explanations, to sincerely express inner feelings, and so on. This would seem to require an ideal form of life in which freedom, equality, and justice had already been attained. At times, Habermas talks as if this were attainable, while at other times he talks as if having this as an ideal will help us to direct the development of discourse in the right direction in our imperfect world.

But the question that we must ask is, even if it were attainable, or even if one were to accept it as a guiding ideal, would this ideal speech situation provide the grounding of ultimate views of the good? In his earlier writings, Habermas writes within the somewhat Marxist tradition of the Frankfurt school, but in his later writings and his espousal of the ideal speech situation, he is more like an old-style European liberal. His approach to the ideal speech situation is value laden. It expresses the democratic ideal of open debate to reach social decisions. This may be an ideal most of us would espouse, but I think we would also recognize that there are many people who would disagree. Christians who feel that the Bible has the truth may welcome open debate as to how best to interpret the Bible, but not open debate to prove God wrong. Similarly, Marxist-Leninist society has the notion of a vanguard party of people with the special skills to lead society, quite contrary to open debate with full equality for all. While I myself believe in the importance of ensuring certain patterns of openness and equality in discussion, it is not clear to me that this ideal speech situation would yield convergence to agreement on the ultimate good, nor is it clear to me that such convergence would be desirable.

Historically, consider the evolution of religion in Europe. There was a time in which each state or principality had its state religion, Catholic or Calvinist or Lutheran. If you were unfortunate enough to believe in other than the sanctioned religion, you found yourself open to censure, censure that was fueled by the conviction of the people who espoused the state religion that they had as "the one true religion." Most of us would see it as a sign of progress when Europe reached a stage of religious tolerance in which people could indeed express differences of religion. But what would happen if there were an ultimate consensus? Can you imagine the society in which it was obtained? It would indeed be stultifying, because if agreement had come about for all right-thinking, symmetrically communicating people as to what was correct, all individual opinion would be rejected as an invalid disturbance of the ultimate consensus.

Ideology

This discussion suggests that ideology will continue to play an important role and that convergence to the one true ideology is unlikely. With this, the time has come to look at ideology in some more detail. For most of us, the search for values is hard to root in fact. But Marxism suggests that it can be so rooted.[2] The classic Marxism of the older Marx is "scientistic" in that it is realist about values. This Marxism offers a value-free science of human society. Values are objective because they are determined by the laws of historical progress offered by this theory. Within this theory, the word ideology is seen as pejorative. Ideology is an infrastructure whereby the ruling class asserts its power. Ideas are seen as epi-phenomena of social and economic conditions, embodying a false consciousness whereby the ruling class can falsely believe in the justice of its position and whereby the proletariat can falsely be convinced of the inevitability of their unprivileged position. What Marx sees as the inevitable results of the dialectic of history is the uncovering of this false consciousness of ideology. Social structures will be seen in the clear light of science, and the proletariat will inevitably come to victory because only the proletariat, deprived as it is, can serve as the interest-free bearer of freedom and in it ultimate virtue will necessarily reside.

But do we really believe that ideology can be made completely explicit? And do we really believe that any class or any human can be without interests? I suggest not. But first, let us look at the criticism internal to Marxism. By the middle of this century, many people who still found much of value in the writings of Marx wanted to find a way of preserving some aspects of

Marxism while recognizing the failures of the predictions of the historic dialectic. Consider the analogy of Christianity at the end of the first century. The early members of the church had based their belief on the imminence of the end of the world. After decades had passed and the world was still with us, a time had come for a reinterpretation of the founding text. And so it is that, with the passing of the first century of Marxism, many people came to reinterpret the text, yielding the so-called "Marxism with a human face," which gives much emphasis to the writings of the young Marx and denies that science is value free. It offers a Marxist critique of the natural sciences as linked with Western capitalist, industrialist societies which support the status quo and technological exploitation. Within this perspective, current conditions might indeed call for revolution, but it is no longer possible to see such revolution as being part of the deterministic unfolding of history. More play is to be given to individual freedom and the determinism of history becomes somewhat muted. The trouble with this is that once one allows science in general to be seen as value laden, the criticism can be turned back on the science of Marxism itself. One may then ask what ideological functions Marxist theory serves in whatever historical circumstances surround it. Once again, we are teetering on the brink of total relativism in our theory of knowledge. But I think we can begin to address this.

We begin by first ridding ideology of its pejorative connotation, accepting that ideology is not to be seen as imposed upon us by history as necessarily bad, but is rather a necessary expression of society. There have to be social schemas to bind the members of a society together, to allow them to live together, even in the simple case of agreeing which side of the road to drive on. This transfer of an ideology from something necessarily repressive to something necessary does not deny that a critique can be made of it. This has some resonance with the history of Freud's notion of the superego. Freud saw the superego as repressive, being the residue from the early interactions with the father of the behaviors that were denied to the child. But in modern psychoanalysis the ego and superego are in a much more constructive dialogue. Another Freudian term for the superego, the ego ideal, comes into use as we see the ego ideal as expressing a representation of society without necessarily being something which unduly cripples the developing human.

The Social and the Individual

To help make the transition to looking at ideology in the more individualistic terms of our schema theory, let me mention one more scholar of sociology, namely Durkheim. In his analysis of religion (Durkheim, 1915), certainly a very important factor in social value systems, he came to reject Marx's one-way explanation of religion in terms of the economic superstructure. Instead he tried to focus attention on the multiple interactions between different social institutions—the religious, the economic, and others—even though one can look at each of these as relatively autonomous. He stressed that the effectiveness of a religion depends in part on the effectiveness of its rituals; and this, in turn, depends on the effectiveness of belief. The social system must be psychologically internalized; a clarion call for schema theory! We need a theory of the internal workings of the symbols of an ideology or religion. We need to study individual beliefs not in isolation, but to see how they reinforce each other as well as seeing their relation to the external world. Evidence for these accumulates from many directions. We have to understand that the justification for a social schema comes not only from the natural world, but also in terms of personal emotions, social cohesion, the social history of conflict and personal relations. We are again in the process of a multilevel dialogue. I now want to explicitly address this notion of an ideology as something outside, but not as something immutable. It is something that the child comes to as an external reality and internalizes to become a member of society, but the way in which the ideology is internalized does not preclude a critique. We are again reminded of Hesse's philosophy of science. In becoming a member of a scientific community, one must learn how to make observations and one must learn the techniques of theory building. But in learning these, one comes to gain a sufficiently rich expressive language so that one not only understands what observations and theory came to coherence in the work of earlier scientists; one also can provide a critique of extant theory in terms of one's own observations and eventually change theories.

In chapter 2, we used language as a focus for our discussion of the interaction between the acquisition of individual schemas, that process by which the individual comes to be a member of society, and the social schemas, the patterns of coherence or commonality which seem to define social structures as an external reality. In trying to balance the social and the individual, we reject the classical Marxist view that the external reality provides the infrastructure determinative of individual action. We are closer to what certain Marxists call an interpretive theory, a theory that merges the horizons of the ideology as a fixed external reality with the network of individuals' beliefs. One interpretive theorist is William Connolly, formerly of the University of Massachusetts, who contrasts social roles as external entities with what an agent must know to hold these roles. To be a worker, to accept a contract, you have to understand what a contract is, you have to understand what it would mean to break the contract. To have this knowledge, Connolly asserts, is to have the potential to criticize the roles, to reject the roles. But in his book, *Appearance and Reality in Politics*, Connolly (1981) still seems to view these roles as necessarily imposed and bad, as suggested by his view of someone who accepts a particular role in society:

> The individual . . . can respond affirmatively to all their pressures by gradually adopting the beliefs appropriate to his role. The others [the other elements of the critique] can simply be allowed to fade away. And by the time the worker reaches this situation, he need no longer see himself, at least consciously as either a coward or a victim . . . and [will believe] that young dissidents who challenge these views have had little experience in the real world.

Ideology here is still seen as bad, while accepting a role in present society is seen as necessarily suppressing elements of critique. And yet, can't we perhaps see an ideology, a set of roles within society, as providing an evolving model for social relations and an important tool for mental economy?

I remember a conversation when I was an undergraduate about the ideal democracy. We were speculating about computer technology and the idea that with computers in the home, people would be able to dial in each night and vote their decisions about the pressing issues of the day. One friend destroyed the whole picture very nicely by just saying "Decisions, decisions, decisions." None of us has the time to learn enough to make all the decisions necessary for our complex society to work. We want the checks and balances of democracy so that we can provide some high level input as to the overall nature of decision making. We should be able to use the pragmatic criterion to decide when things have got out of hand and a change of government is called for. But to be involved in making a decision every moment as to the course of society could only overload us. Similarly, I would suggest that it is a tool of mental economy to view social roles and social schemas, including ideologies, as necessary and as necessarily provisional. This does not mean that one cannot form a critique. In fact, if schemas are provisional, as our whole discussion of science and schemas has suggested, then the more reflective amongst us *will* form a critique.

Let us see the bad and the good of what happens when one forms a critique. When the ideology is repressive, someone who dares to form a critique may be subject to political or religious persecution, trouble holding a job, ostracism, exposure to ridicule, troubled social relationships, and so on. Of course, people may certainly see situations in which, having formed a critique that leads to all these adverse consequences, they may nonetheless have a view of reality that leads them to persevere in the critique. If the pragmatic criterion reveals a discrepancy between two things, it does not specify which thing is wrong. If someone learns that his view of society is out of joint with society itself, that can mean that society is messed up, or that the person is messed up, or both! But in some cases, we have a body of individual experience, or we belong to a subgroup in society that has sufficient cohesion, and we persist, we become rebels, the critique continues. In other cases, the costs, as in the case of Connolly's "acquiescent worker," prove too much, and we come in due time to repress these elements of critique. Then either these elements fade away and we become a happy member of society or they continue as a festering sore, which we cannot do anything about, but which yield a certain level of unhappiness. But of course, if one does not view ideology as *necessarily* a bad thing, one can see that the critique can proceed in other ways. One may indeed be persuaded that the critique is misguided; or one may see that, given the current state of society, the benefits of change are outweighed by the likely costs of trying to bring that change about. On the other hand, critiques can contribute to the evolution of social institutions. Society does change, and while some of these changes can be seen as accompanied by great costs, others can be seen as progressing in a relatively comfortable way.

Schemas are our set of operational means for changing things.

As we noted before, the language in which we exercise our critique is inevitably value laden. As a child, we start with the interests of easing discomfort, or of getting enough to eat and drink. As our schemas grow, they always interact with schemas embodying personal comfort and social interaction. So as we grow up within a society, by the time we have reached an age and level of education in which we could provide a critique, much of the vocabulary in which the critique is expressed is already laden with the values of that society. That is inevitable, even if the language is not as biased as Orwell's NEWSPEAK. Even if we do see things wrong with the society, we may become overwhelmed with the complexity of society and find it easier to live with the status quo than to come up with an implementable plan of action.

From a cynical point of view, one might say that an ideology succeeds to the extent that it excludes conflicting schemas from consideration by making their reality appear untenable. But no matter how holistic, no matter how integrated, no matter how encompassing an ideology or other symbol system may be, the network is open to other experience and this may either lead to evolution of the system or to rejection of the system. One historical example that is very close to me is that of Czechoslovakia. When I first visited there in 1964, I found perhaps the most prosocialist country in the Eastern bloc. But the Czechs, while in favor of socialism, were convinced that no single vanguard party knew the best way to implement socialism. They were dedicated to democracy, not as something that would lead to a rejection of socialism, but as something that would conduce to hermeneutic dialogue within the overarching ideology of socialism. So it was very exciting to see the Prague Spring of '68 and, under Dubcek's guidance, to see the country coming to what seemed to be its sought-for process of dialogue. But in October 1968, the Soviet tanks rolled into Prague, and this noble experiment in dialogue was destroyed. This crushing of their experiment by the Soviet

Union led many of the ardent Czech socialists to a critique of the situation so all encompassing that they were unable to hold to a socialist framework. In some people, the experience of a socialism which uses force led to a continued battle, for others it led to the ultimate critique of a nation, emigration. The irony is that this critique is based on a vivid internalization of the promise and its destruction and that, in the 1980s, a Czech has to be overy thirty to remember. As Milan Kundera has observed, the tragedy of Eastern Europe is the loss of memory.

Relativism

Are we to see an ideology like the rules of the road, a set of purely conventional choices that allows us to live together; or can we hope to find an ideology that embodies absolute values which will let us condemn slavery or Nazi genocide? I doubt that we can find such an absolute position. It was only in 1983 that the northern and southern branches of the Presbyterian church of the United States were reunited, healing a split caused 100 years ago because people were able to read the teachings of Jesus Christ as either supporting slavery or rejecting it, depending on which part of the country they lived in. The process of hermeneutic dialogue or schema change leads at best to a process of learning in terms of current criteria, rather than absolute progress to the right text, the right science, the absolute ideology. It was in this vein that we qualified the Piagetian view of schemas with Freud's view that schema change could lead to conflict as well as to coherence. New schemas that give us further control of some new aspect of our experience may lead to tensions and conflict with what we already know within other contexts. And so we come back to our critique of Habermas, that interests are inevitable and cannot be circumvented to yield an ideal speech situation. In finding it impossible to imagine a society in which people are devoid of interests, I am reminded of

Ambrose Bierce's definition, in *The Devil's Dictionary*, of an egotist as "a person of low taste, more interested in himself than in me." Interests are something that *other* people have.

Our schema theory has led us to see social systems as necessarily provisional, as necessarily open to critique through the process whereby a society, no matter how well established, is continually internalized by new members of the community who bring individual experience to bear. And because of that experience, they may in time generate critiques. But, in line with our hermeneutic approach, we have to stress that this schema theory is itself just one more perspective. Schema theory is itself falsifiable. It is underdetermined by data. It does claim to provide understanding of human beings, but only through its own perspective. We have tried to use this perspective to create an epistemology from which we can analyze and compare other perspectives. However, and this is the point, we do not have to claim that this perspective is absolute. The perspective of schema theory owes its appeal in part to the increasing impact of computer science, the relative success of cognitive science, and the general suspicion in much of our culture of nonnatural entities, such as souls, in defining persons. Given the whole thrust of this volume, to admit to such perspectivism is not self-defeating. It recognizes the need for making prejudgments, and it allows for criteria in terms of which this perspective can be compared with others. In this spirit, we offer schema theory, not to provide recipes for absolute value, as Habermas might claim to do with the ideal speech situation, or as Marxism might do with the historical dialectic, but rather to help us understand the diversity of value systems and to help us analyze how they interact and how they may change. Where a philosopher sees beliefs as primary in an account of the mind, we seek to understand beliefs in terms of schemas instantiated in the brain.

Chapter 5 Freedom

WHEN THE natural sciences were the sciences of matter and energy, the separation of mind from body had a plausibility that is lost now that we can study the processing of information in scientific terms. Computer science, communication theory, cognitive science—these not only have tremendous technological implications, but also help change our very concept of what it is to be human. Schema theory, as we have offered it here, attempts an information-processing account of the human mind, and this account is enlivened by the possibility of artificial intelligence. However, claims for increasing machine intelligence need not entail that all the predicates we would apply to human intelligence and social behavior will necessarily apply to machines, any more than the success of physiology leads us to ask the pressing question, "Will machines have indigestion?" This is because, as we saw in chapter 1, much of what is human about our intelligence depends upon our having human bodies and living as members of human society. Thus, our concern has not been to try to reduce the human to the status of our current machines, or of those we can envision for the near future, but rather to see how schema theory, our chosen line of cognitive science, can be developed both as science and as philosophy, to suggest ways in which the science of information, this new dimension in science, can be engaged in a dialogue with the humanities.

It was such a dialogue that Mary Hesse and I shared in preparation for the Gifford Lectures. As Mary and I tried to plan the lectures that we gave in Edinburgh in November 1983 we were in the ideal speech situation, to use Habermas' term, for we were in a position of intellectual equality and neither of us had power over the other. Yet, for all of this, we certainly did not reach consensus. It may help to see what we did and did not agree upon. Certainly we merged two horizons. I came to the enterprise with my thoughts about schema theory with its stress on mechanisms in the head or the computer. Mary Hesse came to our work with a concern for philosophy of science and a particular stress on the sociology of knowledge. Our collaboration saw the integration within one epistemological perspective of an understanding of both the individual and of the social.

The Voluntarist View of Human Freedom

Despite the integration of the individual and the social in our thinking, differences still remained. For Mary, God was still to be seen as a transcendent reality, where I would see a God-schema as a social construct. And turning to the theme of this chapter, we differed about free will. I hold the position of what we called the "decisionist" who claims that to have free will is simply to act on the basis of one's schemas, and sees the term "freedom" as useful at a social level of analysis: we are free if others do not unduly coerce or indoctrinate us. This contrasts with the "voluntarist," as we called Mary's viewpoint, who locates freedom at the physical level—with the free choice of a self transcending physical law. The voluntarist says, "Surely, it is I who choose whom I marry or what I will eat at the restaurant. It cannot be that the current set of schemas and the state of the world can determine my choices, or the probability of those choices. How can you account for my feeling that I have many alternatives before me and that I can choose between these future worlds? How do you account for my sure knowledge that I have many times had to think long and hard to search my very soul before making agonizing choices that have drastically affected not only myself, but other people who are very important to me?" For the voluntarist, then, there must be the notion that human choice can break the chain of causality. Thus, in particular, voluntarism is incompatible with determinism. The decisionist does not either require or reject determinism. For the decisionist, it is simply enough that schema activation, and this includes consciousness, involves processes that occur within spatiotemporal reality. The challenge for the decisionist is to see how these processes, instantiated within the brain, can still give us an experiential life which is diverse and rich in the way that we subjectively believe to be the case.

The voluntarist must reject a notion of complete predictability, such as that of LaPlace, because it seems inconsistent with her notion of choice. The voluntarist feels that choices can be made that will actually change the course of physical events. The crux for a voluntarist who wishes to come up with a hermeneutic dialogue with the findings of science is to see how this freedom of choice could be compatible with—though not determined by—the laws of nature. It is suggested that the answer may be found in quantum mechanics, which is not deterministic; rather, the future is only predictable at the level of probabilities, and these uncertainties, these probabilities, can affect the macroworld. A random movement of an electron can be amplified to throw a switch one way or the other. Membrane noise in a neuron can determine whether or not a neuron reaches threshold and this can affect some mental state. The voluntarist will then accept the quantum-mechanical view of physics as being our current "best bet" about how the world works, and then build upon the claim that quantum mechanics only determines a *set* of possible future states of the world to claim that the unique individual can transcend spatiotemporal reality to choose within that ensemble of possibilities. Notice, however, that this is not a crude notion which reduces human choice to the random tossing of a quantum-mechanical coin. We are not free, says the voluntarist, because we are random, but rather because randomness leaves us room in which to make individual choices. In other words, this element of choice is compatible with, but goes beyond, physical causality.

The voluntarist thus regards the mind/brain as essentially an open system, in terms of which some events are simply uncaused at any physical level and so needs some supplementary account of free will in terms of entities that are not physical, not in space and time. Now, in a classic form of voluntarism, namely dualism, the Cartesians (or Eccles, whom we discussed in chapter 3) postulate a soul that interacts with the brain in physically inexplicable ways, and so changes the sequence of physical causality. How then does the soul affect the physical world if it is separated from the physical laws of nature? This concept seems to break the notion of the unity of the person, because even a voluntarist may be impressed by the progress made by cognitive scientists in showing how much of the features of intentionality can indeed be explained within information processing terms, and so need not be detached from the action of schemas. We have this tension, then, between the person and between the unfolding of schemas within space and time. In chapter 1, we conceded that it would perhaps be presumptuous to think that our physical knowledge, as currently expressed, is such that we could hope to infer from it a complete theory of the human mind. Instead, we suggested the utility of a notion of two-way reduction in which there would be a dialectic between two different realms of science which would lead to their continuing process of integration. From this perspective, one is to see the language of persons as useful for encapsulating meaningful patterns of what our brains can do, but not as describing a distinct reality, even though person-talk might describe many phenomena distinct from those we can *now* understand in terms of schemas, in terms of brains. With this, my overarching philosophical perspective was that of a continuing mutual adaptation and unifying of these levels of discourse.

However, in our progress towards Edinburgh, Mary Hesse came to reject this form of two-way reduction. She argued that the essence of the human was an emergent property, compatible with physical causality, but not entailed by it or any subpersonal development of it.

An emergent property is one that we can observe in a complex structure that we cannot observe in the isolated pieces. With a statistical averaging assumption, we can understand many emergent properties of a fluid or a magnet in terms of a theory of the underlying "pieces," a physical theory of molecular interactions. Now, the decisionist would hold that mental properties are of the same kind, though this has certainly not been proved. He would argue that, though it may require some developments in the underlying theory, we can explain all phenomena in naturalistic terms. By contrast, we may speak of the *strong* view of an emergent property as one that emerges as a new property for systems of sufficient complexity, which cannot in any sense be explained in terms of lower-level interactions.

Hesse takes the strong view of the person as an emergent property. She cannot foresee a future development of the physical or the brain sciences which would come to explain what it was to be a person in terms of the interaction of schemas or neurons. It would require a new analysis of the specifically human which had to be seen as something that could emerge within a sufficiently complex mind-brain system, but which could never be deduced from the working of that system, no matter what the change of our laws of physics or physiology might be. In this way, there could be downward causation from this emergent structure, the mind-body complex. This "mind-body complex"—not a disembodied soul acting from outside upon the body—would constitute the person, and would affect the way neurons interact. Choice, decision, responsibility, praise, and blame—all these predicates—would then only be explicable in terms of this emergent complex of the human person and could not, even in principle, be reduced to theories of the brain.

With this, the voluntarist has reached a point of definite disagreement with the decisionist. The voluntarist is looking for some transcendent essence of the human, even if it is an emergent essence, while the decisionist sees human personality as arising from our complexity within space and time as conditioned by our human bodies and our interactions within society. The voluntarist claims that the phenomenological stories that people tell about their freedom are not to be reinterpreted into some other scientific story that changes their meaning. This position says there is a "me," a someone beyond whatever is happening in my brain, that makes the decisions, that agonizes over difficult choices, that has human experiences. If you have followed Professor Hesse this far (and the argument is developed in more subtle detail in chapter 5 of our book *The Construction of Reality*), and if you find this emergent, voluntarist view resonant with your view of your experience as an individual, then we would claim that you are forced to a dramatic conclusion: If you are a secularist who is also a voluntarist, then your secularism has been "impaired" in that you can no longer hold that all human reality is within space and time. If you are a secularist/voluntarist, you may come to identify these transcendent entities beyond space and time as persons. But at this stage the theist would step in and suggest that the concept of person as a transcendent entity is parasitic upon the more radical religious affirmation of the transcendent reality of God and that it is relations with God that constitute this transcendental personhood of human beings.

The Decisionist View of Human Freedom

Turning from this viewpoint, I want to argue that the decisionist can give a coherent view of human freedom and can integrate this into a secular world view that does not require transcendent principles, whether those of God or those of Habermas' ideal speech situation. Needless to say, in the short space that remains, it will not be possible for me to completely replace 4,000 years of the Judeo-Christian tradition! At most, I can offer a number of promissory notes. To preface the decisionist case, I would recall the individualistic or liberal conception of freedom (Partridge, 1967) that goes back, for example, to John Stuart Mill's *Treatise on Liberty*. On this view, one is free to the extent that one is free from coercion, able to set one's own goals, and to choose between alternatives. One is free to the extent that one is not compelled to act as one would not choose to act, whether that compulsion comes from the will of another human or from the state. In analyzing this freedom, the liberal view sees that there are different forms of coercion. There can be the direct form of coercion in terms of commands or prohibitions backed up by force or superior power; but there are also the indirect forms of coercion that mold the range of alternatives that are even considered, whether through propaganda or through education. These latter are forms of coercion that do not restrict the choice already made, but rather restrict the wherewithal to consider alternatives and to make then a deliberate or informed choice. Literacy and freedom of the press are thus parts of this liberal conception of freedom because they provide the ability to express diverse viewpoints so that people can make their choices within an understanding of a rich repertoire of alternatives.

The decisionist view, in so far as it is a view of freedom (as distinct from an approach to cognitive science), is really a version of this liberal view. For the decisionist, you are free if you act in the absence of external constraints or physiological debility, but rather act in terms of your internal schemas, part inherited, part acquired, part socialized. Where one might be a liberal and still a voluntarist in holding that the "you" who makes these free choices is a transcendent person, the decisionist instead sees the "you" that decides as constituted by the holistic net of schema interactions within your brain.

Let us recall, from chapter 1, that people carry in their heads what appears to them to be a total model of the universe; and what they experience must be assimilated to their current stock of schemas, though it may also change them. The child or adult is an active participant in his world; when confronted with something new, he tries to "tell a story" to fit the new data, starting from the current stock of schemas. This new story becomes a contribution to the extended stock of schemas that constitutes their personhood within space and time. What they already know provides the equipment, the machinery, through which they incorporate new data. But what is crucial, what we alluded to when we discussed consciousness, is that this highly personal network has a *coherence* and it is this coherence of the network that we call the self. A self seems to be something that develops in a sentient being able to reflect on his experience. Not all of us have equally coherent selves any more than we are equally sensitive to experience, but everyone is conscious of being and having a self. And so it is in this sense that we can find the "you" that makes decisions to be explicable within space and time, within the network of schema interactions that is at the core of the decisionist view of the person and, thus, of the decisionist view of freedom.

What, then, of the voluntarist's sense that one often has to think long and hard before making agonizing choices? It seems to me that one can begin to get some sense of this within the jargon of computer science by invoking the notion of computational complexity. We can be aware of many alternatives that have to be considered before reaching a decision. We may be aware that we have no easy rules, no preset evaluations, which can quickly determine the best alternatives. We may not even be sure what constitutes "best." We may also have some sense of the costs of a wrong decision. A life, a career, or the love of someone dear to us may hang in the balance on the basis of this decision. But the fact that we believe that decisions can be made through the working out in space and time of our schemas, which are but imperfect reflections of natural and social complexity, does not say that there is any quick shortcut to computation. The fact that a computer program is deterministic, and I've said that the schema view is not necessarily deterministic, does not deny that it may take many hours or even days of computing before the computer will deliver an adequate answer. And so we, aware of these complexities, may find ourselves forced to make a decision knowing that we have imperfect data, knowing that we have too little time, knowing that the outcome of the decision is of immense importance to us. And here, it seems to me, is a sense of "thinking long and hard to make agonizing choices" that requires nothing transcendent for its understanding. In fact, we are back to our brief discussion of Gödel's theorem in chapter 1, where we characterized much of the decision making that we engage in, whether or not it is conscious, as being a time-limited process of analogical reasoning, rather than the completely elaborated, certain deduction from consistent axioms.

Going further, the voluntarist has a rich vocabulary with which to talk about the phenomenology of our everyday experience, including such terms as "feeling guilt," "accepting responsibility," and "deserving praise or blame." Let me attempt a decisionist account of a couple of these in a rather improvised way as an invitation to further explore whether the language of schema theory can indeed bring us to some sort of closure in giving an account of subjective experience. To analyze "feeling guilt," we can go back to our discussion of Freud and the superego and then reflect it forward into schema-theoretic terms: schemas can embody our sense that there are socially sanctioned courses of action, but also an awareness that our own schemas are not simply devices for following the social consensus. We may then be aware that in a particular situation the latter may lead us to make choices which are discordant with those we believe society or our parents would ask of us. In some cases, we are able to call upon our stock of schemas to rationalize our course of action, but in other cases we must confess to a weakness or impulse that caused us to act in a way that we cannot choose to defend. It is this notion of indefensibility that might provide the core of a decisionist account of guilt. All those who have ever been on a diet will know what I am talking about! In the same way we can talk about notions of responsibility. In fact, the sense of reponsibility may be more pressing for the decisionist than for the believer in a soul: if you believe that there is a soul that provides a pipeline to God, then responsibility reduces to whether or not you open the communication channel. From a decisionist viewpoint, we may say that you are a responsible member of the society not so much if, in a standard situation, you emit the stock response, but rather if, finding yourself in a new and complex situation, you can nonetheless marshal your schemas in such a way that you act in some way that others, with time and leisure to reflect, would agree to be appropriate.

Punishment at any time is something that is historically rooted. We may provide a critique of it, either by trying to ground it in transcendental principles or in terms of social mechanisms. Having done that, we may either wish to preserve a tradition, or we may change the form of punishment in terms of its effectiveness either as a deterrent or as an example to others. A term like "punishment" thus comes to be seen as shorthand for a variety of complex situations that we can analyze at great length. But what then of a phrase like "he deserves punishment"? A decisionist might say that it is shorthand for some expression like: "Here is a person who in many situations has shown himself capable of evaluating certain alternatives. Here is a situation in which the person did not choose to avoid some grievous situation. We have no way of explaining how that choice was made, save to see it as a failure of certain accepted principles of choice on that occasion. For that reason, the person will be punished." Clearly, to the extent that we come to accept such shorthand, to that extent will the affective force of many terms in our current moral vocabulary be altered. However, it would be wrong to think that any single decisionist expansion of the above kind could exhaust the meaning that such terms have as part of our person reality. As was argued in chapter 2, much of our language is metaphorical, evocative without being exhausted by any simple literal paraphrase.

When we talk about diminished responsibility in a court case, we are simply trying to evaluate whether the mind of the defendant is such that he has the wherewithal, given a situation, to come to a "responsible" decision. Much of the law—as in distinguishing manslaughter from murder, or arguing for diminished responsibility—can be seen as turning on the attempt by lawyers to construct a decisionist theory of the person. The impact of the law is to be judged, not on the basis of a transcendent entity (though such axioms as "all men are created equal" are in some sense the Enlightenment extracts from a transcendental Christian view of the soul), but in terms of an analysis of human decision making, worked out as we would say through our schemas, to give an account of the role of deterrence and punishment, relating social goals to those of the individual.

The Marxist View of Human Freedom

The liberal view sees freedom as the absence of coercion or indoctrination by another. But for Karl Marx, freedom is that which gives man the ability to exercise conscious rational control over both the natural environment and social forces. The key to Marx's argument was that if we can come to see social relations not as part of an external natural reality, but rather as being man-made, then poverty, for example, can be seen as the result of a man-made social order and thus as an enforced limitation of freedom. This view of freedom comes not only through the minimization of external constraints, but also involves the creation of optimal conditions for everyone's fullest personal development. This seems to be an admirable extension of the liberal view; however, when coupled with faith in the total truth of a particular world view, it has had tragic results.

In his 1844 article "On the Jewish Question" (translated in McLellan [1977]), Marx condemned the liberal view of freedom as basing the right of man to liberty not on the association of man with man, but on the separation of man from man. He argued that the liberal view made every man see in other men, not the realization of his own freedom, but the barrier to it. Marx saw no positive value in privacy, but sought to subordinate the private to the public sphere, in what Walicki calls a kind of democratic totalitarianism.[1] By contrast, liberal freedom does involve the notion of a private sphere in which no state, not even the most democratic, has the right to interfere. Marx saw politics under capitalism, then, as only serving the egoistic interests of bourgeois man, and the Russian populists who followed him concluded from this that political freedom was merely a deception and that it was not worth fighting for its introduction. This was not Marx's view, but it is true that he failed to support freedom and privacy for its own sake, as a principle.

Marx did not see society as having an obligation to its present citizens. He saw individual freedom as incompatible with the historical development which was to yield planned, conscious control of the socioeconomic infrastructure. The legitimation of the social order was to be provided by the inner logic of history, not by the will of the electorate. His followers concluded that the logic of history gave them a mandate to exercise power in the name of its final goal. The lack of concern with humans as individuals led to totalitarianism.

As we have seen, Marx held that, to be free, man must be able to exercise conscious rational control over his natural environment and social forces. For the first, freedom requires maximizing productive forces, not reversing industrialization in the name of preindustrial harmony. For the second, freedom means a conscious shaping by men of the social conditions of their existence to eliminate the power of alienated, reified social forces. Freedom thus conceived is inseparable from rational predictability and opposed to the irrationality of chance. Market mechanisms are to be replaced by "production by freely associated men, consciously regulated by them in accordance with a settled plan." We leave aside questions of whether such full consciousness can be attained, or how a plan is to be finally settled, and instead stress that Marx really believed that nationalizing the means of production would end the separation of men's capacities as a species from a man's capacity as an individual and would result in a universal "species man" with "the entire wealth of previous development encompassed within each individual."

But this is a utopian vision. He did not foresee the possibility that, even in a socialist welfare state, individuals might not want to be "liberated" from the egoism and particularism that were incompatible with what Marx held to be man's "generic essence." His disciples went beyond Marx to conclude that it would be necessary to force men to realize their true generic essence through indoctrination and by physical elimination of the "unteachable."

A Secular Schema for Society

Now perhaps it is not presumptuous to warn ourselves that the concurrent emergence of computer technology and cognitive science requires an immense effort of the philosophical imagination if it is not to lead to yet another massive tragedy in human experience. For it is here, I think, that man's self-understanding will be forged for the coming age. This new technology allows us to make massive changes in workplace, in social structure, and in our self image. With what do we structure this endless array of choices? Our struggle in these chapters has been to build upon a schema theory limited to the more scientifically tractable areas of cognitive science as we begin to chart the density and complexity of human moral and emotional experience. Yet I do not think that there is any model of our future society which is *implied* by what we have learned from schema theory.[2] To see why this is so, consider two lessons of schema theory: "Schema theory teaches us that people are inherently different because each has his own stock of schemas." But if we try to infer from this a social view, are we to infer that this view must respect individuality, or are we to see the need for imposing tighter controls so that individuality will not damage the social consensus? "Schema theory tells us that knowledge is provisional." But does this lead us to respect pluralism, or does it rather say that we should restrict inquiry to reduce the risk of destabilizing change?

Given these possibilities, I can offer a secular schema of society compatible with a decisionist, nontranscendental view of the person, but with no claim to certainty that it is "correct." It incorporates the notion that a person may be free if he chooses to follow custom, but not if he has been conditioned to always follow the changing dictates of a ruler. Freedom seems to require not so much that one must always choose to change, but rather the possibility of such choices, and for this the various alternatives must be known (Partridge, 1967). I thus join those who hold the liberal view of freedom in seeking a society in which a wide variety of beliefs are expressed and where there is a considerable diversity of tastes, customs, codes of conduct, and styles of living. I would see this as more likely in a society in which power is widely distributed. My schema, then, is one of a pluralistic, adaptive society which accepts that tradition may contain much wisdom of continuing relevance, yet sees that portions of the tradition are harmful or simply irrelevant in the modern world. Certain schemas—think of those that guide warfare as we enter the nuclear age—were acquired in a sociohistorical context, such that they are inappropriate to be acted upon in a modern context. Even worse, the present context lacks the necessary feedback or social pressure to recalibrate them.

We have repeatedly argued that our knowledge is provisional. Yet there are some secularists who do seek ultimate principles in which to ground human values. We have seen that Habermas thought that an ideal speech situation would lead the participants to a consensus that would define ultimate human values. From a different perspective, E. O. Wilson (1978), in developing the human implications of his sociobiology, has argued that "the good" is whatever evolution shows to be fittest in terms of natural selection. But I would argue, as would the religious believer, that our moral intuitions may require us to resist the latest social consensus or even the ultimate biological consensus and may cause us to feel that we as humans have learned something that cannot be expressed within the blind processes of natural selection and evolutionary trends. Yet, unlike the theist, I resist any notion of *ultimate* principle though I do not deny the value of a set of well-grounded principles to guide our interactions within a particular sociohistorical context. I would see that our critique of society must be an ongoing process rooted in lived experience and cannot see religious belief as providing a touchstone. (The contrary view is eloquently developed by Mary Hesse in *The Construction of Reality*, see especially chapter 11. Some of the present arguments are further developed in chapter 12.)

While I would argue that there *are* doctrinaire forms of religious belief that are definitely dangerous, an enlightened theist like Mary Hesse might respond as follows: The way we read the Bible is part of a hermeneutic process. The Bible was written by men who believed that the end of the world was at hand. Thus, at the end of the first century, as we have seen, people had to learn to read the Bible in a new way, because the temporal world was enduring. When Augustine wrote at the Fall of Rome, he found a way of reading the Bible that gave comfort to men who saw the greatest works of civilization crumbling by offering instead the endurance of the city of God. Aquinas was shaped in great part by writing his *Summa Contra Gentiles*, his attempt to refute the teachings of Islam. Thus there are religious believers who would say that, though the root of human existence is still to be found in man's relation to God, the tradition is not something static. It is a body of wisdom that can and must adapt to the age. When one surveys the history of the Christian religion, one sees that Christian society has often been totalitarian—the Inquisition, and outbreaks of anti-Semitism, and of sectarian violence. There are also places and times in history in which religious belief has allied itself with the fullest expression of human freedom. In the present world, we can perhaps look at the Roman Catholic support for the Liberation Theology of Latin America (a support that is not shared by the hierarchy) or the Solidarity labor movement in Poland. One can argue that it is an evolving sense of social justice, rather than religion, that is driving these movements. Witness the long-standing support of the Church in Latin America, until recently, for the privileged rather than the oppressed. What makes one a secularist, rather than a religious believer, is that one cannot accept a transcendent view of salvation, but rather finds oneself engaged in an evolving critique of society as lived within this present world.

I am not in a position to prove that the decisionist secular viewpoint is superior to a religious view. I can simply say that through my own body of experience I have come to find myself with a schema network that is secular and that, as I expose it to a dialogue with Mary Hesse and with the writings of various religious traditions, I have been able to account for certain aspects of the human condition that some people would find to require a transcendent view of the person. Following Freud's notion of projection, but not his myth of the primal horde, I see ways in which a group of people within space and time might come to the historical belief in God, without saying that a God-schema is itself a manifestation of a transcendent reality. This does not deny at all that many people still find belief in God to be beneficent. One of the great challenges of theology is that of theodicy which is, essentially, the question of why, if God is so good, the world is such a place of travail. In particular, why is there evil? The story of the Fall from the Garden of Eden is a very powerful story: man was created in a certain blessed state, but because he has free will, he was able to fall from grace; but now, through God's goodness, he has the chance to redeem himself, so that no matter what the vicissitudes of the current situation, whether the Holocaust comes, whether Rome falls, there is still meaning beyond our personal death, even beyond the death of our entire village, or the extinction of our civilization. For many people, it is the power of this story that makes them reject the secular view which accepts that, just as the dinosaurs became extinct, so too will humans become extinct and nothing of the human spirit will endure.

My view, however, is that there are no ultimate axioms, no canonical books. Rather, we are continually reacting, changing, updating our views in response to historical processes. In this present world, we cannot look exclusively to the grounding of ancient principles, but rather must grapple with the new challenges of nuclear war, abortion, civil rights, the plight of refugees, damage to the ecology, and the changing realities of family, friends, and workplace. What becomes of the specifically human in such a secular view? All I can say is that we must adapt current data about personal experience as part of what is to be explained without denying that such experience may be open to change. Human values are not grounded in God or any other ultimate principle. Rather, we continually start anew from our current values and subject them to a critique which is itself history laden. The question, "What are the grounds of human values?" may be put in perspective by the apparently frivolous question, "What are the grounds of kangaroos?" There is no essence of the ideal animal that justifies the existence of the kangaroo. Rather, we use neo-Darwinian theory to understand the evolution of the kangaroo within a rich historical and geographical context. It seems to me that this evolutionary view must also apply to human societies and to the religious texts that form an integral part of the culture of those societies. The religious texts offer many resonant metaphors that seem to help us in trying to chart the complexity of our experience, using such words as "evil," "salvation," "destiny." Are we to give up these words? Perhaps not, but I think we must defuse many of the metaphors that infuse them with their current meaning. We can no longer see evil as some absolute concept, whether it be embodied within the Devil or whether it be some principle with which God and the God-fearing are locked in eternal struggle. There are institutions and actions and people that we may agree are evil, but no matter how much we may hate them in our recent history, I do not believe we can come to understand them by an appeal to some absolute essence. Rather, we will accept the existence of evil in its many occurrences as something to be fought, and we will see the betterment of the human lot as something to be sought. But we will deny again that there are absolute criteria for a salvation to be sought in a God-reality outside human history.

Our problem is that we have not yet learned to be fully secular. Our society is pluralist, but *how* are we to be pluralist? The fact that we agree that a rich, free society is one that tolerates many viewpoints does not say that we condone unbounded tolerance. I think that we are in a society that for all its pluralism has agreed that there are certain "don'ts"—genocide, slavery, and kicking dogs. Our society may be in transition with respect to our views on homosexuality, but abortion is still a matter of agonizing debate for which I see no grounds for immediate consensus one way or the other. Given this situation, what we need is not a set of axioms that all members of society should hold in totality to yield a consensus on all these values, but rather a society in which there are accepted patterns of communication which allow for moral discourse while not necessarily determining its outcome; a society that allows us to come to accept and live with the provisional nature of our personal and social knowledge.

However far the process of critique and consensus may go in a society, the construction and interaction of schemas will continue to create surprises. Evolutionary processes create new species which by their very existence provide new ecological niches and possibilities for the evolution of yet further species. Just as there is no single solution to the problem of adaptation in biological evolution, so there is no unique solution to the search for human values in a changing and complex world. It is an old cliche that "the truth shall set you free," but all too often tyrannies are founded by those who would force others to accept their truth. Schema theory does not offer the final truth, only the freedom to learn.

Notes

Chapter 1. The Embodiment of Mind

1. For another perspective on cognitive science, see Pylyshyn (1983) and the attendant commentaries in the same volume. My own commentary suggests that Pylyshyn's view is typical of that of a majority of cognitive scientists in emphasizing the roles of the computer and of linguistics and paying too little attention to neuroscience.
2. See Chomsky (1957) for his original theory of syntactic structures; (1965) for the basic description of his theory of transformational grammar; and (1981) for his recent theory of *Government and Binding,* which embodies major changes in his thinking about syntax. Lyons (1977) gives a useful exposition of Chomsky's life and work.
3. A brief discussion of this historical background, together with a compact proof of Gödel's Incompleteness Theorem, is provided in the last chapter of Arbib (1964). A second edition is promised "soon." See also Nagel and Newman (1958).
4. Piaget was a prolific author. For an appreciation of his work read *The Construction of Reality in the Child* (1954), while the more recent *Biology and Knowledge* (1971) gives an overview of his "genetic epistemology" which develops an embryological metaphor for the growth

of a human's, and of human, knowledge. Boden (1979) gives an appreciation of Piaget's work from the perspective of a philosopher knowledgeable about artificial intelligence. Since many of Piaget's conclusions are based on informal observations of child behavior, much work has been done on psychological testing of his theory under more rigorous conditions. Forman (1982) presents an excellent sample of such work, with an insightful epilogue by the editor.

5. Our basic studies of pattern-recognition mechanisms subserved by the interactions of retina, tectum and pretectum may be found in Lara, Arbib and Cromarty (1982) and Cervantes, Lara and Arbib (1985). Two alternative approaches to modeling detour behavior are offered by Arbib and House (1983, 1985) and Lara et al. (1984). For a view of how classical ethology relates to modern neurophysiological studies, see Ewert (1980).
6. The classic text on ethology is Tinbergen (1951). An autobiographical perspective is given by Lorenz (1981), while Ewert (1980) discusses the neural mechanisms underlying animal behavior.
7. The concept of "virtual finger" was introduced by Arbib, Iberall and Lyons (1985) who illustrated it in the task of picking up a coffee mug. They also discuss the neural implementation of such schemas. Iberall, Bingham and Arbib (1985) develop the notion of "opposition space" to structure the brain schemas for hand movement. Overviews of the work are provided by Iberall and Lyons (1984) and by Arbib (to appear a).
8. A general perspective on schema-theoretic models of language is provided by *From Schema Theory to Language* by Arbib, Conklin and Hill (in press).

Chapter 2. Language

1. For an exposition of theorems relating the internal and external specifications of automata, see section V, "From External to Internal Descriptions" of Arbib (1973).
2. See, e.g., Trevarthen (1982). This paper "rests on the belief that the human brain has systems which integrate interpersonal and practical aims together from the first few months after birth. Evidence for this belief comes principally from close observation of how young infants behave in interaction with other persons and with objects, but it also has support from studies of human brain growth."
3. A good exposition of Piaget's concept of reflective abstraction is

given in his chapters in Beth and Piaget (1966). An appreciation and critique of this approach is given in Arbib (to appear c), where I argue that Piaget pays insufficient attention to the role of social structures (including formal instruction) in the child's construction of logic and mathematics.

4. Lightfoot (1982) argues strongly for the Chomskian position on the innateness of linguistic universals which is set forth in Chomsky (1968). Jill and Peter de Villiers (1978) provide a good overview of observations on how children acquire their native tongue. Wanner and Gleitman (1982) collect papers which offer a lively range of data and theories. Part III of Arbib, Conklin and Hill (in press) provides a "neo-Piagetian computational viewpoint."
5. The idea of language as metaphor is developed by Arbib and Hesse (in press), building on ideas in Hesse (1980). Indhurkya (1985) offers a computational theory of language comprehension which requires no special mechanisms to handle metaphor.
6. See, for example, the provocatively titled essay "The Unreasonable Effectiveness of Mathematics in the Natural Sciences" in Wigner (1967).

Chapter 3. Freud

1. Eccles (1977) gives an excellent undergraduate-level introduction to current neurophysiology, with special emphasis on synaptic transmission, the mechanism of communication between brain cells. Eccles received the Nobel Prize for his contributions to this subject. The last chapter reviews evidence on split-brain patients and then presents the theory of the liaison brain that I am criticising here. More recently, Eccles (1982) has suggested that the liaison brain is located in the supplementary motor area. For a more materialistic account of the role of this brain region, see Goldberg (1985).
2. See Miller (1956) for "the magical number 7 plus or minus 2," and the critique by Neisser (1976).
3. See Weiskrantz et al. (1974), while Weiskrantz (1974) reviews experiments on monkeys designed to probe this phenomenon. Perhaps the most intriguingly titled paper on this topic is that of Humphrey (1970).
4. For an account of the work of Jackson in the perspective of nineteenth-century studies of both evolution and the localization of functions in the brain, see chapter 6 of Young (1970).
5. In Arbib (1981b), I have carried out an explicit analysis of a cooper-

ative computation algorithm (in this case, one for computing optic flow [Prager and Arbib, 1983]) to show it could be improved by such a Jacksonian mechanism. The algorithm does not function well near edges of objects, but another algorithm can be designed to detect this "breakdown" and hypothesize edges; this new information about edge-hypotheses can then be used to refine the original algorithm.

6. In the remainder of this chapter, I will refer to a number of works of Freud by their year of initial publication. In most cases, I will not list explicit references in the bibliography, since readers can usually find paperback editions published by Penguin or Norton, or may turn to the Standard Edition of *The Complete Psychological Works of Sigmund Freud*, under the general editorship of James Strachey in collaboration with Anna Freud, assisted by Alix Strachey and Alan Tyson. For the reader seeking an overview of Freud's life and work, I recommend Wollheim (1971).
7. See Freud (1953). Arbib and Caplan (1979) provide a schema-theoretic perspective on neural mechanisms of language which includes a historical review of Freud's work in relation to that of other nineteenth-century neurologists, including Broca and Wernicke.
8. The English translation of the *Project for a Scientific Psychology* was published as pp. 347–445 of Freud (1954). Wollheim (1971) carefully assesses the role of the *Project* in Freud's development. Pribram and Gill (1976) offer an assessment of the *Project* in relation to Pribram's expertise in neuroscience and Gill's in psychoanalysis. There has recently been a resurgence of interest in attempts to link psychoanalysis and neuroscience. Two recent examples—neither of which refers to the work of Pribram and Gill—are Reiser (1984) and Winson (1985).
9. See Note 8.

Chapter 4. Schemas and Social Systems

1. For a range of readings in hermeneutics, see Schleiermacher (1958), Dilthey (1961), Weber (1980), Collingwood (1946), Gadamer (1976) and McCarthy (1981). The order is chronological in terms of the original work, rather than of the cited publication or translation.
2. For English translations of the work of Karl Marx referred to in what follows, see McLellan (1977). A brief introduction to Marx's thought is offered by Singer (1980).

Chapter 5. Freedom

1. My further analysis of Marx's views on freedom has been greatly influenced by the article by Andrzej Walicki (1983) on "Marx and Freedom."
2. A clear example of this may be seen by contrasting the views of two of the fundamental contributors to cybernetics, the ur-discipline of cognitive science. Where Wiener (1964) espouses an essentially decisionist position, MacKay (1982) argues for the compatability of scientific knowledge and Christian faith.

References

Arbib, M. A.

1964 *Brains, Machines and Mathematics.* McGraw-Hill.

1972 *The Metaphorical Brain: An Introduction to Cybernetics as Artificial Intelligence and Brain Theory.* Wiley Interscience.

1973 Automata Theory in the Context of Theoretical Neurophysiology. In *Foundations of Mathematical Biology,* vol. 3. *Supercellular Systems,* ed. R. Rosen. Academic Press, 193–282.

1975 Cybernetics After 25 Years: A Personal View of System Theory and Brain Theory. *IEEE Transactions on Systems, Man, and Cybernetics, SMC*-5:359-363.

1981a Perceptual Structures and Distributed Motor Control. In *Handbook of Physiology—The Nervous System II. Motor Control,* ed. V. B. Brooks. American Physiological Society, 1449–1480.

1981b Visuomotor Coordination: From Neural Nets to Schema Theory. *Cognition and Brain Theory* 4:23–39.

To appear a Schemas for Temporal Organization of behavior. *Human Neurobiology.*

To appear b Brain Theory and Cooperative Computation. *Human Neurobiology.*

To appear c A Piagetian Perspective on the Construction of Logic. *Synthese.*

Arbib, M. A., and D. Caplan

1979 Neurolinguistics Must Be Computational. *Behavioral and Brain Sciences* 2:449–483.

Arbib, M. A., J. Conklin, and J. C. Hill

1986 In press. *From Schema Theory to Language.* Oxford University Press.

Arbib, M. A., and M. B. Hesse

1986 In press. *The Construction of Reality.*

Arbib, M. A., and D. H. House

1983 Depth and Detours: Towards Neural Models. In *Proceedings of the Second Workshop on Visuomotor Coordination in Frog and Toad: Models and Experiments,* ed. R. Lara and M. A. Arbib. COINS Technical Report 83–19. University of Massachusetts/Amherst.

1986 In press. Depth and Detours: Towards Neural Models. In *Vision, Brain and Cooperative Computation,* ed. M. A. Arbib and A. R. Hanson. Bradford Books, MIT Press.

Arbib, M. A., T. Iberall, and D. Lyons

1985 Coordinated Control Programs for Movements of the Hand. In *Hand Function and the Neocortex,* ed. A. W. Goodwin and I. Darian-Smith. *Experimental Brain Research Supplement* 10: 111–129.

Arbib, M. A., K. J. Overton, and D. T. Lawton

1984 Perceptual Systems for Robots. *Interdisciplinary Science Reviews* 9, 1: 31–46.

Bartlett, F. C.

1932 *Remembering.* Cambridge University Press.

Beth, E. W., and J. Piaget

1966 *Mathematical Epistemology and Psychology*. Translated from the French by W. Mays. Reidel.

Boden, M. A.

1979 *Piaget*. Fontana Paperbacks.

Borges, J. L.

1975 Of Exactitude in Science. In *A Universal History of Infamy*. Penguin Books, 131.

Brainerd, C. J.

1978 The Stage Question in Cognitive-Developmental Theory. *Behavioral and Brain Sciences* 1: 173–213.

Bresnan, J.

1978 A Realistic Transformational Grammar. In *Linguistic Theory and Psychological Reality*, ed. M. Halle, J. Bresnan, and G. A. Miller. MIT Press.

Burks, A. W.

1980 Enumerative Induction vs. Eliminative Induction. In *Applications of Inductive Logic*, ed. L. J. Cohen and M. Hesse. Oxford University Press, 172–189.

Cervantes, F., R. Lara, and M. A. Arbib

1985 A Neural Model of Interactions Subserving Prey-Predator Discrimination and Size Preference in Anurans. *Journal of Theoretical Biology* 113: 117–152.

Chomsky, N.

1957 *Syntactic Structures*. Mouton and Co.

1965 *Aspects of the Theory of Syntax*. MIT Press.

1968 Recent Contributions to the Theory of Innate Ideas. In *Boston Studies in the Philosophy of Science*. Reidel, 3: 81–107.

1981 *Lectures on Government and Binding*. Foris.

Coetzee, J. M.

1984 How I Learned about America—and Africa—in Texas. *New York Times Book Review*, 15 April, p. 9.

Collingwood, R. G.

1946 *The Idea of History*. Oxford University Press.

Conklin, J.

1983 Salience as a Heuristic in Planning Text Generation. Ph.D. Diss., Department of Computer and Information Science, University of Massachusetts at Amherst.

Connolly, W. E.

1981 *Appearance and Reality in Politics*. Cambridge University Press.

Craik, K. J. W.

1943 *The Nature of Explanation*. Cambridge University Press.

de Villiers, J. G., and P. A. de Villiers

1978 *Language Acquisition*. Harvard University Press.

Dilthey, W.

1961 *Pattern and Meaning in History: Thoughts on History and Society*, edited with an introduction by H. P. Rickman. George Allen and Unwin.

Durkheim, E.

1915 *The Elementary Forms of Religious Life*. Translated from the French by Joseph Ward Swain. George Allen and Unwin.

Eccles, J. C.

1977 *The Understanding of the Brain*. 2d ed. McGraw-Hill.

1982 The Initiation of Voluntary Movements by the Supplementary Motor Area. *Archiv für Psychiatrie und Nervenkrankheiten* 231: 423–441.

Ewert, J.-P.

1980 *Neuroethology: An Introduction to the Neurophysiological Fundamentals of Behavior*. Springer-Verlag.

Forman, G. E., ed.

1982 *Action and Thought: From Sensorimotor Schemes to Symbolic Operations*. Academic Press.

Freud, S.

1953 *On Aphasia*. Translated from the German. Image.

1954 *The Origins of Psychoanalysis, Letters to Wilhelm Fliess, Draft and Notes: 1887–1902*. Basic Books.

Gadamer, H.-G.

1976 *Philosophical Hermeneutics*. Translated and edited by David E. Linge. University of California Press.

Gödel, K.

1931 On Formally Undecidable Propositions of *Principia Mathematica* and Related Systems. English translation by B. Meltzer, 1962. Basic Books.

Goffman, E.

1974 *Frame Analysis: An Essay on the Organization of Experience*. Harper and Row.

Goldberg, G.

1985 In press. Supplementary Motor Area Structure and Function: Review and Hypotheses. *Behavioral and Brain Sciences*.

Gregory, R. L.

1969 On How So Little Information Controls So Much Behavior. In *Towards a Theoretical Biology, 2: Sketches*, ed. C. H. Waddington. Edinburgh University Press.

Habermas, J.

1971 *Knowledge and Human Interests.* Translated by Jeremy J. Shapiro. Beacon Press.

1979 *Communication and the Evolution of Society.* Translated by Thomas McCarthy. Beacon Press.

Hanson, A. R., E. M. Riseman, J. Griffith, and T. Weymouth

1985 In press. A Methodology for the Development of General Knowledge-Based Systems. *IEEE Transactions on Pattern Analysis and Machine Intelligence.*

Head, H., and G. Holmes

1911 Sensory Disturbances from Cerebral Lesions. *Brain* 34: 102–254.

Hesse, M.

1980 *Revolutions and Reconstructions in the Philosophy of Science.* Indiana University Press.

Hill, J. C.

1983 A Computational Model of Language Acquisition in the Two-Year-Old. *Cognition and Brain Theory* 6: 287–317.

Hill, J. C., and M. A. Arbib

1984 Schemas, Computation and Language Acquisition. *Human Development* 27: 282–296.

Humphrey, N. K.

1970 What the Frog's Eye Tells the Monkey's Brain. *Brain Behavior and Evolution* 2: 324–337.

Iberall, T., G. Bingham, and M. A. Arbib

1985 In press. Opposition Space: A Structuring Concept for the Analysis of Hand Movements. *Experimental Brain Research Supplement.*

Iberall, T., and D. Lyons

1984 Towards Perceptual Robotics. Presented at the *1984 IEEE International Conference on Systems, Man and Cybernetics*.

Indhurkya, B.

1985 A Computational Theory of Metaphor Comprehension and Analogical Reasoning. Ph. D. Diss., Department of Computer and Information Science, University of Massachusetts at Amherst.

Keat, R.

1981 *The Politics of Social Theory: Habermas, Freud and the Critique of Positivism*. Chicago University Press.

Kertesz, A.

1982 Two Case Studies: Broca's and Wernicke's Aphasia. In *Neural Models of Language Processes*, ed. M. A. Arbib, D. Caplan, and J. C. Marshall. Academic Press, 25–44.

Kuhn, T. S.

1962 *The Structure of Scientific Revolutions*. University of Chicago Press.

Küng, H.

1979 *Freud and the Problem of God*. Translated by Edward Quinn. Yale University Press.

Lara, R., and M. A. Arbib

1985 A Model of the Neural Mechanisms Responsible for Pattern Recognition and Stimulus Specific Habituation in Toads. *Biological Cybernetics* 51: 223–237.

Lara, R., M. A. Arbib, and A. S. Cromarty

1982 The Role of the Tectal Column in Facilitation of Amphibian Prey-Catching Behavior: A Neural Model. *Journal of Neuroscience* 2: 521–530.

Lara, R., M. Carmona, F. Daza, and A. Cruz

1984 A Global Model of the Neural Mechanisms Responsible for Visuomotor Coordination in Toads. *Journal of Theoretical Biology* 110: 587–618.

Lieblich, I., and M. A. Arbib

1982 Multiple Representations of Space Underlying Behavior. *Behavioral and Brain Sciences* 5: 627–659.

Lightfoot, D.

1982 *The Language Lottery: Toward a Biology of Grammars.* MIT Press.

Lorenz, K. Z.

1981 *The Foundations of Ethology: The Principal Ideas and Discoveries in Animal Behavior.* Simon and Schuster.

Lucas, J. R.

1961 Minds, Machines and Gödel. *Philosophy* 36: 112–127.

Lyons, J.

1977 *Chomsky.* Rev. ed. Fontana/Collins.

MacKay, D. M.

1966 Cerebral Organization and the Conscious Control of Action. In *Brain and Conscious Experience,* ed. J. C. Eccles. Springer-Verlag, 422–440.

1982 *Science and the Quest for Meaning.* Eerdman.

Matthei, E.

1979 The Acquisition of Prenominal Modifier Sequences: Stalking the Second Green Ball. Ph.D. Diss., Department of Linguistics, University of Massachusetts at Amherst.

McCarthy, T.

1981 *The Critical Theory of Jurgen Habermas.* MIT Press.

McDonald, D. D.

1983 Natural Language Generation as a Computational Problem: An Introduction. In *Computational Models of Discourse*, ed. J. M. Brady and R. C. Berwick. MIT Press, 209–266.

McLellan, D., ed.

1977 *Karl Marx: Selected Writings*. Oxford University Press.

Miller, G. A.

1956 The Magical Number Seven, Plus or Minus Two: Some Limitations on Our Capacity for Processing Information. *Psychological Review* 63: 81–97.

Minsky, M. L.

1975 A Framework for Representing Knowledge. In *The Psychology of Computer Vision*, ed. P. H. Winston. McGraw-Hill, 211–277.

Nagel, E., and J. R. Newman

1958 *Gödel's Proof*. New York University Press.

Neisser, U.

1976 *Cognition and Reality*. Freeman.

Pais, A.

1982 *Subtle is the Lord . . ., The Science and the Life of Albert Einstein*. Oxford University Press.

Partridge, P. H.

1967 Freedom. In *Encyclopedia of Philosophy*. Macmillan, 221–225.

Piaget, J.

1954 *The Construction of Reality in the Child*. Translated from the French by Margaret Cook. Basic Books.

1971 *Biology and Knowledge: An Essay on the Relations between Organic Regulations and Cognitive Processes*. Edinburgh University Press.

Popper, K., and J. C. Eccles

1977 *The Self and Its Brain*. Springer-Verlag.

Prager, J. M., and M. A. Arbib

1983 Computing the Optic Flow: The MATCH Algorithm and Prediction. *Computer Vision, Graphics, and Image Processing* 24: 271–304.

Pribram, K., and M. Gill

1976 *Freud's Project Re-Assessed*. Basic Books.

Pylyshyn, Z. W.

1983 Information Science, Its Roots and Relations as Viewed from the Perspective of Cognitive Science. In *The Study of Information: Interdisciplinary Messages*, ed. F. Machlup and U. Mansfield. Wiley-Interscience.

Ramon y Cajal, S.

1937 *Recollections of My Life*. Translated by E. Horne Craigie with the assistance of Juan Cano. MIT Press.

Reiser, M. F.

1984 *Mind, Brain and Body: Toward a Convergence of Psychoanalysis and Neurobiology*. Basic Books.

Sarraute, N.

1984 *Childhood*. Translated by Barbara Wright in consultation with the author. George Braziller.

Schank, R., and R. Abelson

1977 *Scripts, Plans, Goals, and Understanding: An Inquiry into Human Knowledge Structures*. Lawrence Erlbaum Associates.

Schleiermacher, F.

1958 *On Religion, Speeches to Its Cultured Despisers*. Translated by John Oman. Harper and Row.

Searle, J. R.

1980 Minds, Brains and Programs. *Behavioral and Brain Sciences* 3: 417–457.

Shattuck, R.

1984 Life Before Language (a review of *Childhood* by Nathalie Sarraute). *New York Times Book Review,* 1 April, pp. 1, 31.

Shepherd, G. M.

1979 *The Synaptic Organization of the Brain.* 2d ed. Oxford University Press.

Singer, P.

1980 *Marx.* Oxford University Press.

Sperry, R. W.

1968 Hemisphere Deconnection and Unity of Conscious Awareness. *American Psychologist* 23: 723–733.

Tinbergen, N.

1951 *The Study of Instinct.* Clarendon Press.

Trevarthen, C.

1982 The Primary Motives for Cooperative Understanding. In *Social Cognition: Studies of the Development of Understanding,* ed. G. Butterworth and P. Light. Harvester Press, 77–109.

Walicki, A.

1983 Marx and Freedom. *New York Review of Books,* 24 November, pp. 50–55.

Wanner, E., and L. R. Gleitman, eds.

1982 *Language Acquisition: The State of the Art.* Cambridge University Press.

Weber, M.

1980 *The Interpretation of Social Reality*. Edited and with an introductory essay by J. E. T. Eldridge. Schocken Books.

Weiskrantz, L.

1974 The Interaction between Occipital and Temporal Cortex in Vision: An Overview. In *The Neurosciences Third Study Program*, ed. F. O. Schmitt and F. G. Worden. MIT Press, 189–204.

Weiskrantz, L., E. K. Warrington, M. D. Sanders, and J. Marshall

1974 Visual Capacity in the Hemianopic Field Following a Restricted Occipital Ablation. *Brain* 97: 709–728.

Wiener, N.

1961 *Cybernetics: or Control and Communication in the Animal and the Machine*. 2d ed. MIT Press. The first edition was published in 1948.

1964 *God and Golem, Inc.* MIT Press.

Wigner, E. P.

1967 *Symmetries and Reflections, Scientific Essays*. MIT Press.

Wilson, E. O.

1978 *On Human Nature*. Harvard University Press.

Winson, J.

1985 *Brain and Psyche: The Biology of the Unconscious*. Anchor Press/Doubleday.

Wittgenstein, L.

1958 *Philosophical Investigations*. Translated by G. E. M. Anscombe. 3d ed. Macmillan.

Wollheim, R.

1971 *Freud*. Fontana/Collins.

Young, R. M.

1970 *Mind, Brain and Adaptation in the Nineteenth Century: Cerebral Localization and Its Biological Context from Gall to Ferrier*. Clarendon Press.

Index